Spoken and Written Discourse in Online Interactions

Common patterns of interactions are altered in the digital world and new patterns of communication have emerged, challenging previous notions of what communication actually *is* in the contemporary age. Online configurations of interaction, such as video chats, blogging and social networking practices demand profound rethinking of the categories of linguistic analysis, given the blurring of traditional distinctions between oral and written discourse in digital texts. This volume reconsiders underlying linguistic and semiotic frameworks of analysis of spoken and written discourse in the light of the new paradigms of online communication, in keeping with a multimodal theoretical framework.

Typical modes of online interaction encompass speech, writing, gesture, movement, gaze and social distance. This is nothing new, but here Sindoni asserts that all these modes are integrated in unprecedented ways, enacting new interactional patterns and new systems of interpretation among web users. These "non-verbal" modes have been sidelined by mainstream linguistics, whereas accounting for the complexity of new genres and making sense of their educational impact is high on this volume' s agenda. Sindoni analyzes other new phenomena, ranging from the intimate sphere (i.e., video chats, personal blogs or journals on social networking websites) to the public arena (i.e., global-scale transmission of information and knowledge in public blogs or media sharing communities), shedding light on the rapidly changing global web scenario.

Maria Grazia Sindoni is Assistant Professor of English Language and Translation at the University of Messina, Italy.

Routledge Studies in Multimodality

Edited by Kay L. O'Halloran, National University of Singapore

Spoken and Written Discourse in Online Interactions

A Multimodal Approach

Maria Grazia Sindoni

NEW YORK AND LONDON

First published 2013
by Routledge
711 Third Avenue, New York, NY 10017

Simultaneously published in the UK
by Routledge
2 Park Square, Milton Park, Abingdon, Oxfordshire OX14 4RN

First issued in paperback 2015

*Routledge is an imprint of the Taylor & Francis Group,
an informa business*

Library of Congress Cataloging-in-Publication Data
Sindoni, Maria Grazia, 1974–
    Spoken and written discourse in online interactions : a multimodal approach /
    By Maria Grazia Sindoni.
        pages cm. — (Routledge Studies In Multimodality ; #7)
    Includes bibliographical references and index.
    1. Communication—Technological innovations.   2. Language and languages—
Technological innovations.   3. Mass media—Technological innovations.
4. Online social networks.   5. Social interaction.   I. Title.
    P96.T42.S449 2013
    302.23'1—dc23
    2012050175

ISBN 13: 978-1-138-92285-3 (pbk)
ISBN 13: 978-0-415-52316-5 (hbk)

Typeset in Sabon
by Apex CoVantage, LLC

To Fabio and our "*valigia dei sogni*"

Sedulo curavi humanas actiones non ridere, non lugere, neque detestari sed intelligere.

I have made a ceaseless effort not to ridicule, not to bewail, nor to scorn human actions, but to understand them.

Baruch Spinoza

Contents

Figures

Tables

Appendixes

Acknowledgments

This book could not have been written without the contribution of all my research informants, most of them digital natives, who patiently led me through the intricate paths of online worlds. I have lived in their world for the past few years, and their involvement in my research has been invaluable.

First and foremost, thanks to my family and to Fabio, who has been my first editor. Thank you for pushing me in the right direction.

I am very grateful to Anthony P. Baldry, who was the first to encourage me to develop my initial speculations on digital interaction. His generosity, constant support and scientific guide will not be forgotten.

Very special thanks also go to Gunther Kress, who also put me on the right track at the very beginning of my research project, and to Susan C. Herring, who suggested I broadened my vision of mode-switching. Precious advice has also come from Carey Jewitt and Diane Mavers. Among the many scholars who helped me during the writing of this book, I owe my gratitude to Kay O'Halloran, Paul J. Thibault, Richard Xiao, Mike Scott and Giuseppe Lombardo. I am also indebted to Douglas Biber, who first reviewed my project. His comments and criticisms have helped shape my work into its present form. All the faults in the book, however, are entirely mine!

Thanks for a careful reading of my work to my friends and colleagues, especially to Tony Harris, María Moreno Jaén, Elizabeth Hart and Mariavita Cambria.

I am also indebted to Erika Azzarello, who is a dear friend, an amazing artist, and the author of the artwork and illustrations.

A very special thanks goes to my fantastic editors at Routledge, Felisa Salvago-Keyes and Deepti Agarwal.

Last but not least, I would like to thank all my students, with particular reference to Giusy Donato, who helped me with crucial technicalities and insights that would have been out of reach for a digital non-native like me.

Introduction

Recent debates in the fields of communication, media and social sciences hold that what shapes human interactions is much more than language, that is, speech and writing. The view that communication can be accounted for in purely linguistic terms is today untenable, and a range of different approaches have come to embrace the notion on which multimodal studies have been originally grounded. Human beings can count on a wide and refined repertoire of communicative strategies beyond *verbal language*. Language is only *a fraction of* meaning-making events. However, in the general, epistemological realignment of disciplines proximal to communication and interactional studies, speech and writing as paradigms have been partially sidelined and even considered as vestiges of the logocentric Western tradition and, as such, have lost their appeal for non-linguists. At the same time, powerful traditions, such as those still ingrained in most linguistic approaches, still hold in the present day, but these may be partially debunked by satellite disciplines, which are conversely concerned with non-verbal facets of communication.

Speech and writing are in effect imbued with traces of the recent past, whereas their interplay was at the core of ideologies underlying cultural discourses and underpinning the armoury of colonial empires. Academics are to some extent still in denial, and this is apparent in scientific and cultural thinking—be it philosophical, phenomenological, sociological or linguistic—that may be overriding the dominant Western rhetoric behind research produced under the aegis of mainstream cultural agendas. In other words, discourses regarding speech and writing, in Foucauldian terms, as well as spoken and written discourse, are something we all have to experience and make sense of in society and individual relationships, but they are not neutral and have inscribed their own counter-discourses in their socio-historical development.

How is spoken and written discourse articulated in everyday interaction? And how are they affected by their transplantation into virtual environments, such as those of the web? Digital environments are parallel worlds to the "real world," where we all live, learn and interact. However, today, more than ever, the real and the virtual world are blurring and overlapping in ways that are still in need of descriptive models. The choice to grapple with virtual worlds through the oral/written interplay is a philosophical move to investigate unstable and richly semiotic communicative and interactional events through the paradoxically traditional keyhole of verbal language. However, this book claims that verbal language, instantiated in speech and writing, needs to be analysed in context, even though the context in question is, as in this case, a fragmented, mesmerizing, virtual conglomerate of bits and bytes. The ways in which oral and written discourses—and all that goes with them, in terms of Western epistemologies—is thus scrutinized in web-based environments, and specifically, in different forms of spontaneous interaction, to explore how language, as it is viewed and *described* in linguistics, is linked to other semiotic resources.

This is why the method used in this work is eclectic. Multimodality is an umbrella theory encompassing other theories, applicative tools and heuristics, in order to tame all the complexities that such a myriad of texts yield. Assuming that different web-based texts and genres generate completely different patterns of socialization and interaction, different integration of language and other semiotic resources and diverging conversational strategies, it follows that different methods will be applied, accordingly. Multimodal studies will thus be accompanied by other approaches, such as digital ethnography, conversation analysis and corpus linguistics. All these theories will be harnessed to account for various models of communication that are deployed in the web-based interactions dealt with in this book, and in particular, the videochat, blogs and YouTube multimodal texts.

The chapters follow the order of expanding circles in terms of socialization and identity as they are buttressed by spoken/written language variation: videochat in multiparty client interfaces, blogging practices in social networking websites and commenting activities in media sharing communities.

The overall aim of this book, in keeping with multimodal/intersemiotic theory, is thus to reconsider underlying semiotic frameworks of analysis of spoken and written discourse in the digital age, in the light of the new paradigms of online communication. Common patterns of interactions are altered in the web galaxy and new patterns of communication emerge, challenging previous notions of what communication actually *is* in the contemporary age.

Online configurations of interaction, such as videochats, blogging and social networking practices, demand a profound rethink of the categories of linguistic analysis, given the blurring of traditional distinctions between oral and written discourse in digital texts. Typical modes of online interaction encompass speech, writing, gesture, movement, gaze and social distance in spontaneous web-based interactions. This is nothing new, but this volume claims that all these modes are integrated in unprecedented ways, enacting new interactional patterns and new systems of interpretation among web users. These non-verbal modes have been sidelined by mainstream linguistics, whereas accounting for the complexity of new genres and making sense of their educational impact is high on this volume's agenda. Other new phenomena ranging from the intimate sphere (i.e., chats, personal blogs or personal profiles on social networking) to the public arena (i.e., global-scale transmission of information and knowledge in public blogs or media-sharing communities) will be analysed in the light of the rapidly changing global web scenario.

Celebrators of the internet heralded it as a technology which placed cultural acts in the hands of everyday users, decentralizing speech, publishing, film-making, radio and television broadcasting (Poster 1995). Transcending the limits of the passive reception typical of the broadcasting era with internet interaction was announced by second media age literature as a major breakthrough (see Rheingold 1993 for communitarian approaches and Guattari 1989 for postmodern theories). Second media age literature also claims that face-to-face interaction has been superseded by extended forms of communication, such as the internet. In this view, technology separates the individual from a "natural state" of interaction, that is, face-to-face interaction that encourages other forms of social interaction and integration. The latter is endorsed through the erasure of ethnic and gender differences (Poster 1997). However, others argue that because face-to-face interaction is so important as a means of connecting people in the society of information, the internet has, as a consequence, become all powerful and all pervasive as a means of instantiation of such connections (Holmes 2005). Holmes, in fact, claims that communication environments *frame* individual lives, regardless of individual communicative acts: "the dominant background connections or mediums by which a given group of individuals are socially integrated come to mediate other levels of interaction" (Holmes 2005: 17). Holmes argues that our times are characterized by a mode of communication that is no longer connected to a specific medium. An old aunt's letter on perfumed notepaper has now become an email or a text message. A telephone call between two people has become a newsfeed. A conference lecture has become a videocast, and so on. The consequences are twofold: a feeling of liberation and greater freedom to communicate—but also the establishment of new social bonds. In other words, whilst experiencing a telephone call, we may perceive it *as if* we were involved in face-to-face conversation *even though we are not*. But we may also have a *half-present* face-to-face conversation, as extended forms of communication affect *what* we experience face-to-face and *how* we do so.

Predominant forms of social integration mediate communication in our age. Postman's media ecology has emerged from McLuhan's influential medium theory, which focused on the medium—not as a material channel conveying messages, but rather as determining the nature of the message itself. Whereas McLuhan famously claims that "the medium is the message" (McLuhan 1964), subsequently arguing that "the medium is the massage" (McLuhan and Fiore 1967) and further claiming that anything that can extend the body's senses and capabilities may be defined as a medium,

Postman (1985), on the other hand, views media in terms of environments, shaping and being shaped by the context. In a similar vein, Wesch (2009b) claims that communication media influence the nature of our interactions, ultimately determining our attitudes, opinions and points of view.

However, this discussion is not limited to human-related interaction. We need to broaden our horizons of investigation and move up another "notch." Another level of analysis helps understand more easily that what we are witnessing is a profound bio-anthropological change. As biological organisms, we human beings are mutants. We are currently in the process of altering our strategies of interaction as a result of new tools. In other words, the incidence of mediation (i.e., the presence of technology) in mediated communication has reached extraordinary levels of stratification and complexity that are paving the way for new forms and norms of interaction (Scollon 1998, 2001). The new emergent texts that modern technology brings about allow oral and written modes of communication to be intertwined. This has sometimes also been the case in the past, most notably with the invention of printing.

The history of human communication has been punctuated by micro and macro biological and socio-historical events that amount to changes or fractures, altering the linear course of human history, and signalling predicaments or conjunctures that require the use of new epistemological tools to study and make sense of them. For example, Galileo's telescope brought about consequences in spatial explorations, much like the Cartesian revolution, which forced men to reconsider their position in the world. The development of visual/graphic semiotic systems pinpoints in time how different cultural stages in human history have been articulated and negotiated. The interpretation of such systems is neither optional nor limited to the development of writing systems. Discussing the evolution of mathematical discourse, for example, O'Halloran argues, "the semiotic visual rendition of the mathematics problem permits a re-organization of perceptual reality" (O'Halloran 2005: 38).

Does our society simply require modes of communication that are more complex and sophisticated than those of the past (Kress 2003, 2010)? Or is it that the human body is changing and that the robotic extensions perceivable (e.g., mobile phones, netbooks) are part of the evolution of man's body?

Biological events include the organic development of the phonatory apparatus in Neanderthal Man, or the laryngeal descent and development of an area for the pharynx in Homo Sapiens (Lieberman 1975). Socio-historical events are connected to inter-organism factors, such as the invention of the technology of writing (Ong 1967, 1977, 1982; Zumthor [1983] 1990), or, maybe less conspicuously, of printing. However, temporal arrangements of these events need to be evaluated according to notably different time scales: for example, biological events, which favoured the rise of language, date back 120,000–150,000 years, according to some cognitive scientists, while human language began to develop at a later stage (cf. Tattersall 1995). To further expand this crucible where men were striving to develop language, we may say that whereas Homo Sapiens developed between 30,000 and 50,000 years ago, the first forms of writing date back to 6,000 years ago (Ong 1982).

What have been defined as socio-historical events obviously took place at much later stages in human history. Considering the human being as a biological organism, tool use is defined in classical ethology as: "the use of an external object as a functional extension of mouth or beak, hand or claw, in the attainment of an immediate goal" (H. van Lawick-Goodall and J. van Lawick-Goodall 1970, quoted in Shettleworth 1988: 481). Despite the recent elaboration of new theories for tool use (St. Amant and Horton 2008), it may be assumed that tools influenced the social goals and means by which human beings developed systems of interaction. Even though human interaction is far more sophisticated than that developed by any other living being, a basic definition of what a tool is needs to be made to tackle the profound cognitive and social repercussions that changes in the use of tools are likely to trigger.

In the 1980s, Ong (1977, 1982) discussed the technologies of the *word* in terms of the distinction between orality and literacy, arguing that the invention of writing caused a deep and unavoidable process of change in the human brain, ultimately modifying our capacity for storing information, transmitting and distributing it. What had been previously focused on the immediate context of communication, relating to immediate needs, was later turned, by the technology of writing, into the possibility of analytical evaluations of the context itself, whence the rise of complexity. In other words, oral cultures do not have documents, but memories. People who are bred in oral cultures learn, but *do not study* (Ong 1982).

What has all this to do with systems of communication developed within the domain of what may be defined as part of the evolution of the web? The answer is that at the present stage, an increase in peer interaction has lowered thresholds in the distribution of information, integration of technologies and the development of new systems of human socialization (i.e., social networks, videochats, media sharing communities, etc.). Digital environments allow new ways of interacting with people, sharing, selecting and discarding information according to taste or opinion, in a post- or hyper-modern version of pop culture, where "copy and paste" techniques are the implied and unquestioned rule. Intertextuality is now so dense and saturated that it has become nullified.

Chapter 1 traces the foundation on which the book is based, and specifically it interweaves the many threads that constitute spoken and written textuality. The chapter is divided into three parts: the first section sketches a concise socio-historical outline of oral and literate traditions in different cultures, also taking into account how the spoken/written divide has shaped the world, clarifying how spoken and written discourse is embedded in wider ideologies and discourses, and to what extent they are strategies hidden in the dominant rhetoric of power in societies. The second part is a linguistic account that explains the different facets that set speech and writing apart. Recognizing that discussing speech and writing as if they were distinct and conflicting poles is a mere abstraction, this part tackles each language mode as a "stereotypical" version, to borrow from Biber (1988). Such differences from a linguistic standpoint are thus introduced, with a view to the possible contexts and settings, where spoken-like and written-like characteristics combine to produce fully articulated discourses. The last part of the chapter is devoted to the discussion of how spoken and written discourses are transferred in the digital domain, also exploring the degree and extent of modifications originated within discourses in web-based interactions. This chapter is intended as a general framework for critical reflection and provides a possible reading strategy and interpretative key to the whole book. Even though some issues are not explicitly taken up in other chapters, the theoretical framework built up in Chapter 1 underpins the subsequent assumptions and informs all the research questions addressed.

Chapter 2 considers the videochat as a new form of interaction that makes use of unprecedented combinations of language and other semiotic resources, drawing on a purposely created corpus of more than 300 hours of recorded spontaneous web-based video interactions. Restricting the definition of *mode* to speech and writing, in tune with Halliday's early definition (1978), the chapter introduces a book's seminal descriptive unit of analysis, that is *mode-switching*. Accounting for the alternation of speech and writing in videochats, mode-switching paraphrases the well-established linguistic notion of code-switching and describes the chance that users have, in a videochat, to combine audio/video exchanges with written comments. It is made clear that mode-switching *stricto sensu* occurs when such alternations are used in the same communicative event, synchronically and for specific interactional purposes. Other resources are accounted for, and these include medium-constrained social distance, body movements and their interpretations, the practice of self-looking, the impossibility of reciprocating gaze and the lack of eye contact. All these resources combine and create meaning-making events that are explored using multimodal theories, and qualitative observations are discussed, drawing from conversation analysis, by adapting its theoretical and descriptive labels to web-based video interactions.

Chapter 3 presents a foray into the world of blogging. Blogs used to be commonly viewed as digital textual instantiations of self-representations, be they in the form of private diaries or independent journalism, among others. However, the chapter points to textual and semiotic evidence of transitions towards more communal blogging practices, in particular with regard to the amalgam of blogs with social networking communities. A corpus of blog entries is used in this chapter to gauge *semiotic resource integration* and *language mode variation*. For this purpose, two different corpora have been created, one annotating the relevant semiotic resources, and the other made up of purely verbal datasets. Datasets are analysed following corpus-based lexical studies to account for spoken and written variation, using keyness analysis and discussing the most frequent lexical bundles in the corpora. The chapter discusses corpora findings and proposes the notion of *resource-switching*, which is broader than mode-switching and includes all the semiotic alternations that can be found in multimodal web-based environments, for example videos, photos, pictures, tagging and hyperlinking, etc.

Finally, Chapter 4 presents the last digital platform under discussion in the book, namely YouTube, one of the best-known video sharing communities. To investigate language mode variation, another corpus is used, collecting verbal comments onto one viral video across a two-year time span. Comments are explored using methods from corpus linguistics, and analyses of variation across spoken-like and written-like discourses are carried out applying a corpus-based lexical model. Analyses are focused again on keyness analyses, lexical bundles, semantic preference and semantic prosody in the corpus. Despite the growing amount of commenting activities on YouTube, findings suggest that comments may be clustered under functionally related heuristic groupings. Such functional categorization into semantic groupings for verbal comments is suggested in the chapter, but their interaction and integration with other semiotic resources is explained invoking the notion of a *multimodal relevance maxim*. The latter is a way of adapting well-established models (Grice 1952) to new environments, in order to clarify how verbal and semiotic resources are dynamically related, following criteria that may be justified only adjusting theories that were framed to account for verbal language.

To summarize, language has been placed in the foreground and in the background at the same time in this book. It is in the foreground, in that it has been the main focus of analysis, and it is in the background, in that other semiotic resources have been considered as fully meaningful and communicative. Paradoxical as it might seem, language in its multimodal affordances is a system to reconcile, and make sense of, the many complexities of human communication in the digital arena—at the same time endorsing and contesting logocentrism.

1 Spoken and Written Discourse in the Digital Age

1.1 AN AGENDA FOR SPOKEN AND WRITTEN DISCOURSE

More than a decade ago, while I was writing my PhD thesis on Creole studies, my supervisor warned me that writing is permanent. I needed to be perfectly convinced about what I was embarking on, he argued, "because once you write something, it stays forever." The slightly ominous undertone of his advice was obviously not intended for my future career and, ironically, my vivid memory of his *spoken* words runs exactly counter the old adage *scripta manent, verba volant*.

Do words fly, as they say? Or may they be as permanent, even *more* permanent than written words? The "I have a dream" speech by Martin Luther King in 1963, or the "Blood, sweat and tears" speech by Winston Churchill in 1940 may be two cases in point. Think of the "Quit India" speech by Mahatma Gandhi in 1942, the 2000 "Holocaust speech" by Pope John Paul II at the Israel Holocaust Memorial, the notorious 1998 "I have sinned" speech by Bill Clinton or, going back in time, the "Spanish Armada Speech" by Queen Elizabeth I in 1588. All these renowned speeches were so important during their time, rousing collective feelings or stirring public outcry. However, to counter these observations, it may be added that these speeches, once spoken, have been made permanent by writing. All of us, however, can cite some private occasion when a speech, a conversation, or even just a single utterance, has been indelibly imprinted on our memory.

This chapter addresses the question of spoken and written discourse in the light of the changes brought about by digital media, tackling the question as to whether well-established notions continue into the present age or need to be re-adjusted to fit a fluid and impermanent environment, such as that of the web. However, oral and written discourses are not clear-cut notions and several studies have shown their degree of intersections, addressing specific questions in individual texts that display oral-discourse-like and written-discourse-like features. A survey into these two language modes will thus allow a fuller appreciation of such intersections that are more than ever surrounding us in the digital scenario.

For the purposes of this volume, speech and writing will be analysed in the first part of this chapter in their abstractions, that is, in their "pure" identity, as abstract and conflicting linguistic notions, each with their own sets of features and deploying their characteristics in typical discourses. In accordance with what Halliday claims, when addressing the question of oral and written discourse, we need not think in terms of *external factors*, such as phonology contrasted to orthography, but in terms of the *discourses* originated within the spoken and written domains (Halliday 1987). In other words, a preliminary discussion will investigate conceptual and theoretical versions of spoken and written *discourses*, that is, those forms of textual practices and genres that may be grouped under the labels of "speech" *or* "writing" (e.g., informal conversations vs. novels). This will be followed in subsequent chapters by a presentation of case studies, where abstractions, categories and genres make way for impure, mixed and hybrid *texts* (e.g., any of the above quoted examples *in a given and specific context*).

Speech and writing as abstract categories may be seen as conflicting views of language modes, including a set of different linguistic traits that set them apart. However, if specific discourse instances are examined in context, their boundaries are fuzzy and neat distinctions are no more than mere naïveté. A similar example may be found in the notions of "Standard English," "General American," or other varieties, each of which is nonetheless connected to a community of speakers, geographical settings and contexts of situations. Categorizations, such as those of Standards, are abstractions, and varieties need to be analysed in context, as used by a group of speakers, to yield more detailed and realistic accounts. However, definitions of Standard and varieties are, nonetheless, useful for linguistic analysis or for teaching and learning purposes.

The same may be said about spoken and written discourses: they are functional models for theoretical and heuristic purposes, but individual examples, taken from both highly formalized genres (e.g., academic prose) and mundane occasions (e.g., everyday conversation), further enrich the picture, creating a mosaic of spoken and written texts, each of them with their own distinctive trait or set of traits.

The idea of a cline or a *continuum* to describe the ongoing and ever-changing relationship between the two poles of speech and writing may be in effect complemented or even superseded by the image of a mosaic, highlighting the indexical nature of an imaginary picture containing *all* possible spoken/written textual tesserae. Even more complex is the digital arena, where well-established models are challenged by intersections of technical and socio-semiotic meaning-making resources.

From a multimodal standpoint (Baldry and Thibault 2001, 2006; Kress and van Leeuwen 2001; Kress 2003; Baldry 2004; O'Halloran 2004, 2005; Jewitt 2009), each socio-semiotic resource, such as speech, image, writing, colour, layout, movement, gaze, music, and so on, is intended as a distinctive contribution to the meaning-making process. Socio-semiotic resources constitute the elements of which a text is composed. Any text is the product of these integrations, and understanding how a text works is possible when these resources are unpacked and *interpreted*. Verbal and non-verbal resources are integrated in multimodal texts, and especially so in digital textuality, where complex textual relationships are instantiated by the deployment of multiple resources. In other words, a website may present written text, still and moving images, videos and music. Videos, in turn, integrate spoken and written language (e.g., live interactions plus subtitling). A blog includes a wide array of resources, both verbal and non-verbal, as well as social networking digital texts. All these examples show that digital textuality is replete with a variety of verbal and non-verbal resources. But how do they work together? To what extent are they integrated? Are users responsive to verbal and non-verbal resources and to what degree? Are calamity-howlers right when complaining that the so-called digital natives will regress or are regressing into a pre-literate state? What is digital literacy, and what modifications are to be expected in traditional notions of literacy?

Speech and writing are not simple and neutral language modes used interchangeably according to the requirements of situation or speaker. They are loaded with ideological affordances that speakers, though educated and fully literate, may not be fully aware. The way oral and written discourses originated and developed in different societies in effect sheds light on such affordances and also on the way they interact within ensembles of semiotic resources to fabricate meaning-making events.

This volume investigates verbal aspects of multimodal texts in the digital world under the agenda of spoken and written discourse. Multimodal studies have laid claim to the importance of visual modes in communication, so severely neglected in other fields of studies until a couple of decades ago. Such an approach has greatly enhanced the field of investigation of non-verbal communication from socio-semiotic perspectives. Non-verbal language plays a major role in human communication, but its positioning in language studies is controversial, as everything that is not within the realm of verbal language has traditionally been held to be *non-linguistic*. For example, phenomena such as the design of objects and places, art and advertising were generally not considered as forms of *linguistic* communication (Scollon and Scollon 2009). However, multimodal studies in broad terms redress such a balance, adopting a more comprehensive approach to human communication. Psychology and cognitive sciences are still studying the impact of non-verbal communication from numerous standpoints, whereas linguistics has conventionally excluded it from its domain of enquiry and epistemological terrain.

Multimodal studies have been providing a refreshing new perspective on research on communication and language, but have sometimes occasioned a presumably involuntary backwash: verbal modes, namely spoken and written, have been left somewhat behind, trapped in the complex realignments of language studies and in the old dichotomy of speech and writing. Far from re-presenting a monolithic approach that refers back to exclusivist linguistic models, this volume addresses the question of verbal communication in a multimodal world. Speech and writing are considered as language modes and as semiotic resources at the same time, on a par with image, colour and music.

As images were gaining momentum in the era of visual communication, language, and its meta-languages, were trapped into an epistemological vacuum, in some instances lacking a full and comprehensive insight into contemporary communication. However, verbal language has come back with a vengeance in today's communication, and theoretical reinterpretations are needed to fit traditional linguistic categories into new—or apparently new—textual configurations. Speech and writing are being sent out of the door, but come in through the window.

This chapter, and the volume in general, is an attempt at tackling these issues, trying to restore the balance between traditional linguistic categories, such as those of speech and writing, *and* a wider array of socio-semiotic resources. Its underlying rationale is grounded in the belief that verbal modes of speech and writing are *still* useful heuristic categories for exploring multimodal communication. Far from believing that CMC (computer-mediated communication) makes these categories "useless" (Goddard 2004), as CMC further blurs binary oppositions that have always characterized speech and writing, this book is based on the assumption that CMC needs to be embedded in previous well-established linguistic notions. Much research on CMC, as Herring has pointed out (2013), has thrown the baby out with the bathwater. Whereas binary oppositions are not flexible models for observations, they are nonetheless valuable to build a model made up of discrete, and thus separately analysable paradigms.

In the next two paragraphs, two perspectives will be examined, clarifying two complementary views on speech and writing. The first is a socio-historical account of orality and literature, providing a concise history of speech and writing, seen from the perspective of social sciences, literature and humanities. This is to investigate how these modes are applied for aesthetic and socially codified purposes in different models of societies, in particular contrasting Western and "Third-World" societies. This preliminary introduction lays the foundations for the linguistic perspective treated in the second part of the chapter, and will unearth the ideological implications behind the oral/literate divide. The second paragraph deals with similar questions, but from a linguistic point of view, discussing speech and writing and their reciprocal relationship as they are used in everyday contexts of communication and for common communicative purposes. These two different approaches are complementary in that they both focus on how people use speech and writing in a given context to meet communicative needs. However, literary approaches tended to impinge on how language modes used to be perceived or stigmatized in a given social context, for example attaching values or flaws to their use or presumed misuse. Modern linguistics, as we will see, has redressed the balance, putting emphasis on differences in prac-tices and discourses, thus removing typical stigma attributed to both speech and writing in communities of practice.

A preliminary discussion on orality and literature will pave the way for the contribution provided by linguistics, with the goal of unravelling social and ideological implications of the speech-writing dyad, if any, that is par-ticularly relevant to be addressed in the digital age. Finally, the third part of the chapter will outline how speech and writing are adapted in web-based environments, giving rise to text types that conflate specific features of the oral-discourse and written-discourse genres, outlining the agenda and meth-ods that will be developed in the subsequent chapters.

The discussion on orality and literature that follows, far from being a foray into non-relevant areas of studies, is of seminal importance for the concerns of this volume, as it addresses the question of speech and writing in the context of twentieth-century intellectual movements.

1.2   A SOCIOCULTURAL OUTLINE OF THE HISTORY OF ORALITY AND LITERATURE

Questions about speech, writing and their reciprocal relationship have al-ways been complex, and their relationship is a core notion in explorations on language *tout court* (Vansina [1961] 2006). Theoretical reflections on speech and writing have tended to oppose or complement each other. Speech and writing seem intuitively different, as are the discourses that may be ascribed to each of them. But how do they differ? Are these differences still relevant in the digital age?

According to the Online Etymology Dictionary (www.etymonline.com), text derives from the "wording of anything written," from Old French *texte* (late-fourteenth century), Old North French *tixte* (twelfth century), from Medieval Latin *textus* "the Scriptures, text, treatise," in Late Latin "written account, content, characters used in a document," from classical Latin *tex-tus* "style or texture of a work." The latter means literally "thing woven" from the past participle stem of *texere* "to weave," from the Proto-Indo-European base **tek-* "make." From this very concise and easily available online account, it is possible to get a glimpse of the strong and continuous tradition that links the notion of textuality with *writing* and *production*, interpreted as both fabricating things and interweaving elements. The struc-tural idea was later associated with the concept of writing in Latin, with the adjunct of *style* and the later addition of the notion of written account in Late Latin, which, in turn, became the writing *par excellence*, the Bible in the Medieval Latin period. Etymology is thus a practical tool for unearthing how associations are accrued in a lexical item.

Hence, *text* and *written words* are inextricably linked, as an etymologi-cal analysis shows. Yet a significant number of societies were very far from construing their experience of culture and cultural transmission via written texts.

From the beginning of the twentieth century, a terminological debate and ensuing scholarly discussions on orality and oral literatures gained momentum. The concept itself of "oral literature" sounds rather paradoxical, in that it blends together two apparently conflicting notions, speech (i.e., "oral") and writing (i.e., "literature"). How can they coexist? However, the question is ill defined in these terms, as oral literatures, which do not exclusively pertain to pre-literate societies, share some essential traits with written literatures, if we consider their overall social and cultural significance. Despite the epistemological impossibility of comparing oral with written genres, rituals, traditional ballads, heroic poems and many other contemporary oral genres, such as pop songs, folk, reggae and protest songs, represent culture and traditions of the societies in which they were composed as much as written literature does.

Oral and written literatures are, from this perspective, *not* the two sides of the same coin, as they originate in different genres, and this assumption possibly still proves true in the digital world. Furthermore, their traditions are imbued with different contexts and societies, and, if examples are selected for analysis, their common ground becomes loose and diluted.

With regard to their dissimilarities, mention should be made about their different modes of transmission, which influence the genre's constitutive features. Its *durability* and the degree of preservation of orally transmitted compositions vary accordingly. Additionally, oral literature is highly dependent on context and mainly relies on *performance*. Hence, studies and interpretations need to take performativity into account. But in practice, until relatively recent times, performativity has been disregarded in research literature. Performance, however, is also a seminal feature of web-based phenomena: an example is found in YouTube videos, where video makers are likely to produce performances of various kinds and the same may be said about video bloggers. However, performance as a complex multimodal and multisemiotic event has been neglected until recently in studies on oral literature. Furthermore, instances of video clips (or vlogs) are far from being considered as instantiations of "orality," which is actually coming back with a vengeance in web-based environments, as Ong predicted in the late Seventies (1977).

This is why studies on oral literatures were traditionally biased for most of the twentieth century, as they were based on aesthetic canons thriving in written (and Euro-centred) literatures. Despite its highly complex aesthetics, oral literature used to be considered the product of "inferior societies," and, as such, was deemed as not deserving serious interest from academics. It belonged to the field of ethnologists and folklorists, interested in the fuzzy notions of "tradition" and "folk-lore," disregarding other aspects, such as authorship.

In spite of this, the study of oral cultures is a tangential, but epistemologically relevant, contribution to the development of social sciences, like linguistics, in a broad sense. Unravelling the history of the development of studies on oral societies and artefacts is the first step necessary for probing into the origins of the modern and contemporary thought regarding language modes.

1.2.1   The Romantic Age: A Blossoming Interest into Orality and Literacy

Typical as it was of pre-literate societies, the Romantics appear to have been the first to have taken an interest in primitive cultures and its artefacts. Positing irreconcilable positions for *nature* and *art* and for *man of art* and *man of nature*, actually predates Romanticism. These concepts are believed to have been transitional in some respects, probably flourishing in the period before the actual blossoming of Romanticism in Europe, but paving the way for the eventual cultural transformations that gave rise to it.

Rousseau praised the man of nature, exalting the value of experiences lived by men in natural environments, untrammelled by the bounds and constrictions of society, which is defined by a set of rules and regulations, that each citizen needs to comply with. The Romantic praise of the *primitive*, *natural* and *free*, that were replacing neo-classical canons of *diction*, *appropriateness* and *decorum*, represented a fundamental step towards the appreciation of pre-literate societies, where the primitive (oral) still thrived. The primitive was associated with oral traditions, where it was supposed real and original sources of poetical inspiration were to be found.

From a Romantic standpoint, unlettered people had the privilege of being able to express themselves in *authentic* poetic forms, as they were pristine, not contaminated by the artificial and corrupt nature of modern societies. Rousseau's opposition was carried to extremes, as it perfectly matched the nostalgic aura that permeated the idea of cultural origins during and beyond Romanticism: poetry was believed to dwell in the past, and modern poets were permanently looking for inspiration in irretrievable, distant and ancient times.

Significantly, it was during the Romantic age that a growing interest in the origins of languages arose. Finnegan ([1978] 1992) contends that the emergence of nationalism in Europe, which was a direct consequence of the French Revolution and the Napoleonic wars, was one of the main features of Romanticism, also arguing that "along with this went an emphasis on local origins and languages, accompanied by an enthusiasm for the collection of 'folklore' in various senses—what would now be called 'oral literature' (ballads, folk songs, stories) as well as 'traditional' dances and vernacular languages and 'customs,'" also claiming that "the political and ideological implications of this return to 'origins' are obvious, and the appeal was all the more forceful because of the Romantic stress on the significance of the 'other' and the 'lost,' and the virtue of the 'unlettered' and 'natural' folk both now and in the past" (Finnegan [1978] 1992: 34). Furthermore, she refers to those paradoxes implicit in the Romantic interest in oral art: a central strand of Romanticism was occupied by *tradition* that, as Hobsbawn and Ranger have shown (1983), contributed to the birth and development of nationalist movements that led to political re-adjustments in several European areas. Tradition, in Finnegan's terms, consists of the actual transmission of a people's *lore* or *Volkgeist*, and this collective—oral—*tradition* of cultural values and themes implies an unconscious perpetuation of collective art, which excludes, by definition, the deliberate and intentional artistic act or imaginative creation by an *individual talent*. The assumptions that seem (or seemed) to provide the foundations of folklore or ethnology are the postulation of art as a collective product, that is not disembodied but, on the contrary, is highly situational and representative of a stage of development of any society. Finnegan stresses that one of the many paradoxical elements in Romanticism lies in the fundamental and privileged role attributed to artists—seen as solitary, aloof, detached from any real context, and creating artistic forms that are personal, subjective and produced from their own individual perspective.

However, great emphasis was placed on the value of collective and oral art, which was believed to be the counterpart of artistry created by a solitary artist—who was, significantly enough, assumed to be a *writer*. Through folk popular art, people were entitled to go back in time and natural spontaneous literary production was associated with *natural* national identity.

It was at this point that a differentiation between the various identities of art was starting to circulate more systematically, in a short time span becoming part of the intellectual climate of the Romantic and post-Romantic age. In fact, a set of binary oppositions was created and internalised, and orality and literature were considered as opposites until relatively recent times, at least for most of the twentieth century. Literature was an allegedly artistic upshot that sprang from an artist's individual work: it was exclusively written, and authorship gained more and more importance during those years of intellectual fervour. Furthermore, written literature was deemed to be highly refined, as its fixed and unchangeable permanence was guaranteed firstly by the practice of writing itself, and, secondly, by the author's personality, which was there to remain as an everlasting signature for future generations. Permanence, durability and stability were thenceforth considered as constitutive features to the more refined arts, with ensuing traditionalist and conservative views on culture.

Conversely, oral compositions were conceived to be performed and, for this reason, many different versions of the same ballad, song or tale circulated in oral cultures. The repetitive motifs and themes were in contrast to *originality* that was valued as a mark of individuality and distinctiveness. Furthermore, oral literature was believed to have been handed down from one generation to another, without the personal intervention of an individual genius, thus emphasizing the collective, natural and spontaneous element at the expense of the individual, artistic and artificial.

It was taken for granted that the communal was, by definition, unwritten, and Claude Lévi-Strauss contributed to such a binary systematization of cultures with his well-known distinction between *la pensée sauvage* and *la pensée cultivée et domestiquée*.

Many subsequent scholars, however, especially in the field of postcolonial studies, have argued that the theoretical or ideological approach of folklorists, ethnologists and even anthropologists was biased and not based on empirical data. For example, Okpewho (1992) warns that at the turn of the century a new interest in oral cultures was directly related to the experience of *colonialism*. As imperialism and the conquest of new lands and peoples gathered momentum, the study of the "other" as a text became a common practice (Okpewho 1979, 1992). Somewhat prophetically, Lévi-Strauss was also of the opinion that with the advent of writing, human relationships become less and less based on global experience, but rather they stem from a written reconstruction of them. As another well-known example, we can cite Said, who argues, "the Orient studied was a textual universe by and large; the impact of the Orient was made through books and manuscripts, not, as in the impress of Greece and Renaissance, through mimetic artefacts like sculpture and pottery. Even the rapport between the Orientalist and the Orient was textual, [. . .] rarely were Orientalists interested in anything except proving the validity of these musty 'truths' by applying them, without great success, to uncomprehending, hence degenerate, natives" ([1978] 1995: 52).

The interest in *other* cultures was fostered by the determination to conquer and dominate culturally. Moreover, the supremacy of the technologies of writing gave a boost to textual cultures, which extolled their artistic products and justified their very existence at the expense of *other* cultures which, in many cases, had less textual impact, as they relied entirely on speech.

But how could the feared *other*, seen as a stranger, foreigner and uncanny presence, become a *text*? Frazer was one of those scholars who first gathered oral materials from Africa, involving a large number of researchers. They collected data, analysed texts from African tribes and also examined some aspects related to performance. Unfortunately, their approach was limited by their textual education. They were not trained to recognize typical oral features; they were neither able to judge the materials they were examining, nor to distance themselves from European—textual—canons and aesthetics. They were concerned with the *evolution*, as it were, of human history and were thus eager to discern patterns of similarities in different cultures.

The problem was that their rough summaries of oral narratives were merely concentrating on plot, story and thematic features, thus failing to recognize other fundamental elements. Furthermore, the evolutionist scholars were persuaded that oral texts were exclusively the product of a community, in other words, a collective art work in which no specific or individual contribution could be expected. However, comparative readings of the numerous versions of an oral composition may in effect shed light on the individual contribution of a performer in any single performance/event, as has been done in much recent research literature.

Diffusionist studies were a consequence of a more general movement in culture and specifically in the flourishing field of linguistics. The Brothers Grimm, for instance, found patterns of similarities in what they called *families of languages*, thus setting the theory of evolution's founding principles of Indo-European languages. The underlying assumption was that each language is a derivation from a "superior" source. Ideas of hierarchies of languages were reinforced by the Grimms' theories, and as a consequence, a hierarchy of cultures was established as imperialism held sway. However biased, studies on oral cultures were the first steps towards a better understanding of supposed primitive cultures.

The sociological approach to oral and folk literature tended to focus on the role played by society and on the relationship between literature and society. Other approaches had previously overlooked some fundamental social aspects that were at work during an oral performance, as they were focused on literary aspects—motifs or themes that were recurring in many versions of songs or hymns, for instance—or analysed recurring patterns in different societies in order to establish regularities or fragmentations. Sociology of literature has placed more emphasis on the role of literature within society. The postulation of two models of societies, as described by Durkheim, has profoundly influenced the positions adopted by sociologists and anthropologists with regard to oral cultures for over a century. The one society is, in Durkheim's terms, the *primitive*, *preliterate* and *oral*, all in all opposed to *modern society* that is characterized by impersonal relationships, the extensive use of writing, and its industrial and mechanized organization.

This assumption has considerably impinged on studies of oral literatures, but has also been contested in that it postulates the existence of two fixed kinds of societies with antipodal characteristics: as a consequence, each allegedly produces its own kind of literature, according to its constitutive features, and satisfying divergent functional needs and aesthetic values. However, field work and empirical data have shown that this is not the case, because oral literature is not produced only in preliterate and primitive societies, but also appears in many other contexts, and thrives in the so-called industrial or modern societies.

1.2.2 From the Homeric Question Onwards: Ideologies and Epistemologies

Mention should be made of the *Homeric question*, involving issues regarding the creation and transmission of the Homeric texts, which began and developed in the nineteenth century. These debates awakened academic interest in questions related to orality, but they also brought into being one of the most significant inquiries into what were and still are symbolically considered as the founding masterpieces of the European culture. Furthermore, interest in typical aspects of oral literature emerged in research literature, such as the role of the *narrator* or *performer*, the event of the *performance* itself and other related questions, such as the individual, personal and creative contribution of the performer. This cast doubts on the presumed communal and organic nature of oral literature.

The most influential contributors were Milman Parry and Albert B. Lord, who, while studying the *Iliad* and the *Odyssey*, discovered that the composition of these works had been influenced by methods distinctive of oral tradition. A long lasting prejudice about oral composition was therefore brushed aside. The idea of an original text and of a primary source from which all was derived was replaced by a more dynamic view: each performance was original in Lord's terms, and was old and new at the same time, because it simultaneously represented a combination of popular wisdom, literary complexities and aesthetic values, but was also the expression of a single epic poet's/poet-singer's creativity, bringing together these strands to shape a new artistic and performative ensemble.

Seen from this perspective, oral literature acquired a special value in postcolonial contexts: it was no longer the product of a primitive and preliterate society, as had been previously assumed by a biased Romantic-derived approach, as it incorporated high refinement, artistry and craftsmanship. Furthermore, it thrived also in industrial and technologically developed societies (McLuhan 1962, 1964), in the form of protest songs or street ballads throughout the twentieth century. Oral artefacts flourish in all societies and epitomize personal and collective artistic needs and forms of expression and have today also been adopted in, and adapted to, web-based environments.

Yet, language modes actually and initially refer to different ways of grappling with the experience of learning and living in the world. In his analysis of the Greek philosophers on the issue, Havelock likewise argues that writing has radically transformed human thought processing, thus enabling a different type of knowledge storing, no longer based on accumulation and memorization, but grounded on abstract categorizations of the world, brought to the fore by the development of the alphabet (Havelock 1963, 1982, 1986).

In a similar vein, Ong (1977) claims that not only did writing change the relationships among human beings, but also radically transformed the organization and structuring of human thought. Human thought had supposedly been developing from a previous oral phase, founded on repetitive knowledge, which drew upon collectively stored memory. The advent of writing paved the way for *sustained and analytic thought*, structured in wider and more complex units, and not following paratactic and formulaic associations (as in the case of the *Iliad* and *Odyssey*), but privileging hypotactic, segmental and personal verbal arrangements of inner states of mind and experiences (Ong 1982).

From this point of view, writing stands for a supposed *absence*, or lack of unmediated meaning, absence of dialogue and co-negotiated verbal exchange. The practice of writing assumes the presence of implicit readers, whom the author will never meet or know. From the real reader's point of view, a text is experienced in solitary fashion. However, the practice of modern reading is hardly comparable to the collective participation in oral performances, where audience reactions and emotions were predicted by the teller/performer, who manipulated responses to good effect. *Voice* stands for living *presence*, encouraging interactivity and empowering human relationships.

Contemporary intellectuals have reacted against this sense of presence of the word and its implicit reductionism, inverting the terms of orality and literacy. Barthes, for instance, argues that the graphic-visual dimension of writing is antecedent to oral communication. Writing cannot be reduced to graphic signs translating sounds, but is permanent, set against the evanescence of voice. Derrida himself deconstructs the "metaphysics of presence" and "logocentrism-phonocentrism," claiming that "by a slow movement whose necessity is hardly perceptible, everything that for at least some twenty centuries tended toward and finally succeeded in being gathered under the name of language is beginning to let itself be transferred to, or at least summarized under, *the name of writing*" (Derrida 1998: 6, italics mine). His idea is that writing comprehends language, hence asserting the primacy of the former and denying it is a supplement of the *phone*. He deconstructs the Saussurean linguistics that invested speech with absolute primacy and that downgraded writing to a referential and derivative role (Derrida 1967, [1967] 1972, 1998).

Other scholars agree that writing has transformed human thought and social relations, but without going so far as Derrida. Goody (2000), for instance, argues the case for the *power of the written word*, which allows cultures that possess writing to dominate purely oral ones, through the accumulation of knowledge about the world. Political reflections also inform other studies, such as those by Saakana, who contends, "Europe's claim to civilization is based upon its destruction of another nation's civilization" (1996: 19). The cultural plunder perpetrated by Europeans paved the way for what Saakana calls the *falsification of history*, basically grounded on the possession of the technology of writing.

The alleged superiority of those who owned this technology was thus mainly exhibited by those who had the power to rule the world, even though it is believed in some quarters that the ancient Greek/European cultural legacy was the result of their cultural and material plunder of Eastern and African wealth and technological knowledge. For example, Bernal's highly controversial book (1987) maintains that during the nineteenth century, much philological effort was invested in purging classical Greece of its cultural affiliations with Egypt and Semitic areas, to postulate "Aryan" origins of Europe, create an image of "pure" European ancestry. Bernal's work ignited a debate in the academic community and whereas some scholars lamented the lack of empirical evidence in his research, others adopted his view in broad terms. Said, for example, argued that images of European authority were buttressed in the constructions of rituals and traditions and when pre-modern societies were beginning to fray, under the pressures of governing domestic and overseas territories, "the ruling élites of Europe felt the clear need to project their power backwards in time, giving it a history and legitimacy that only tradition and longevity could impart" (Said 1993: 16).

From this perspective, the conventional opposition between *oral-primitive* and *written-modern* cultures needs to be reconsidered. In the contemporary scenario, made up of web-based and networked worlds, ideologies that underpin the digital divide are well worth investigating. Oral and written discourses are not only and simply two linguistic modes or channels of communication, but derive from different ideologies and are ingrained into systems of cultural domination. Globalization is instantiated in technological and semiotic resources that produce standardized systems of interactions, as is the case of videochats, blogging practices, social networking and media sharing communities. They apparently erase all cultural differences instantiated by the oral/written divide. But how do these systems change the way in which oral and written discourse are realized in socio-semiotic and multimodal web-based environments? This is another general question addressed in this book.

In the following subsection, some reflections on how linguistics has come to the fore in the exploration of the oral/written *continuum* will pave the way for the chapter's second section, devoted to an outline of speech and writing as two different language modes. Finally, the question as to whether they have been transformed, and to what extent, by the web will be addressed in the third and final section, along with an outline of the book's methodological agenda.

1.2.3   Transitions: From Orality and Literacy to Speech and Writing

Relatively recent field work into orality and literacy has been more firmly based on empirical data and has shown that some preliminary assumptions were misguided; orality and writing consistently overlap, making it difficult and often pointless to distinguish and separate them. Tannen, for instance, points out, "literate tradition does not replace oral. Rather, when literacy is introduced, the two are superimposed upon and intertwined with each other. Similarly, no individual is either 'oral' or 'literate.' Rather, people use devices associated with both traditions in various settings" (1982b: 3). She also contends that oral tradition has been traditionally associated with family and in-group connections, whereas literacy is mostly linked to more formal institutions, such as school.

In recent times, technology has doubtlessly influenced communication, and many scholars predicted in the Eighties a return of orality, with the consequent loss of prestige, even usefulness, of literacy. Ong, for example, makes a good case for *secondary orality*: the electronic age has brought us into the age of a new orality that, however, is highly dependent on writing and literate culture for its existence. McLuhan maintains that literate societies lack essential qualities, which are found in oral-based communities, like spontaneity and warmth. The loss of these qualities accounts for impersonal and detached social relations in industrial societies (1964).

Lakoff also makes the point that "the oral medium is considered as more valid and intelligible as a form of communication than the written, and [. . .] the reasons for this are not mere decline of education, or mental sloppiness, but are rooted in technological progress" (1982: 240). Lakoff claims that orality is so resilient that, in the long run, literacy will be superfluous, and, in the end, disappear. Concerns about language modes affordances also led to different approaches that could countervail previous biases in the question and allow more balanced positions in the domains of education and language policies.

To counter the dangers of the spoken/written dyad, Lakoff proposes a classification into *spontaneous* vs. *nonspontaneous* discourse, which, in her opinion, is more likely to describe the interplay between orality and writing. However, oral-based (i.e., conveying a fake spontaneity) devices in writing require a careful reading, and Lakoff warns against the dangers of forgetting the ultimate non-spontaneity implicit in writing.

Chafe admits that differences between writing and speaking are not related to the medium. In his analysis of Seneca, an oral Iroquois language spoken in western New York State, he has found that, despite its lack of written tradition, language satisfies *every* need of the Seneca community, including rituality and religion. He concludes that "oral literature has a kind of permanence analogous to that of written language, and that the reciter of oral literature is, like a writer, detached from direct personal involvement" (Chafe 1982: 52).

To summarize, orality and writing need not be considered as mutually exclusive modes of communication, but, on the contrary, as ontologically related, both interrogating the epistemological stance and slipping into the epistemological terrain of the other.

1.3 SPEECH AND WRITING: A LINGUISTIC PERSPECTIVE

Here, I will analyse speech and writing in their abstract versions in order to build a grid for analysis for subsequent chapters. As argued previously, the cultural preference for spoken or written discourse has affected the cultural development of different societies, as their use in context is loaded with social and political implications. Issues of literacy in developing countries, on the one hand, and a stigmatized use of spoken discourse in certain communities, on the other, are issues debated in language planning and language policy backgrounds. This includes the more recent question of digital literacy, which is of special relevance for the concerns of this book.

Oral and written discourses are neither as neutral nor abstract as they might apparently seem: they are ideologically saturated and awareness of their full linguistic and ideological implications informs my overall argument. Discourses are in effect ingrained in the society where they come from and are shaped according to social, political and cultural guiding principles of a community of speakers: what happens when these communities are geographically displaced and culturally fragmented? Are such discourses changed in the process, and, if they are, to what extent?

Initially, speech existed long before writing. Speech is less easily traced back to its origins, whereas writing is a better attested technology, even though it has developed different methods to represent sounds graphically. Why has writing been called a *technology*? For several reasons: it is neither as *natural* nor *spontaneous* as speech, it needs implements and supports to be produced (e.g., ink, quill, parchment, stone, paper, pen, keyboard), but more significantly, it needs a script, that is a commonly shared system of representing sounds or meanings. Without a script, any system of writing is impossible.

Writing can represent meanings either through pictorial representations (i.e., logographic scripts) or through sounds (syllabic or alphabetic scripts). The syllabic script pairs a syllable (usually a consonant plus a vowel) with a single symbol, whereas the alphabetic script pairs a single sound with a single symbol.

Although it is not possible to give a date to the first writing, we may cite three different cultures that inaugurated the technology of writing at different times and places and that made use of various technical implements and systems, namely Sumerian (c. 3,000–2,500 BC), Egyptian (c. 3,000 BC) and Chinese (c. 1,500 BC). No system is pure and pristine: all of them present different degrees of integration and blend with other systems and scripts.

For example, the English language was scripted basically using the Roman alphabet. The Etruscan script became the basis for a runic alphabet, but some runes were added to represent Germanic sounds that were needed to implement the full system of sounds, for example the sound *thorn* /Þ/ used to represent the initial /th/, as in *thick*. Until the mid-fifth century, there is evidence that the system of writing was runic, but the Roman alphabet later replaced it. However, Old English included sounds that were not represented in the Roman alphabet, such as the sound *ash* /æ/, made up of the Latin vowels /a/ plus /e/ plus /ð/ (*eth*), as in *than*. However, runes and Old English symbols were gradually abandoned, thanks to borrowings from Norman-French and also Middle English internal modifications. The introduction of print led to other long-lasting changes, such as those introduced by William Caxton, for example the substitution of *thorn*, /Þ/, by /t/ plus /h/, that is <th>.

Among the main landmarks in writing, we may cite the emergence of codex (c. AD 400), the introduction of spacing between words (c. AD 600–800), the Gutenberg Bible (1452) and the Caxton press at Westminster (1476).

But how and when did the transition from print *per se* to print culture take place? According to Baron (2013), print culture took more than 200 years to be developed and includes a set of attitudes, for example the improvements brought about by language standardization, when two books came to be exactly two *copies*, thus laying the foundations for proper print culture that lasted throughout the twentieth century. Presuppositions of print culture consist of:

1. the *durability* of a text (e.g., a text can be read and re-read and also annotated via glosses or *marginalia*, giving rise to the opportunity of adding a personal note to texts);
2. *new conditions of reading* (e.g., silent and individual reading, expansion of private and public libraries, attitudes and opinions associated with the act of reading);
3. *value of books* (books are written, revised, and edited before being published to justify their social and monetary value).

The latter point elucidates the care that is taken to produce the *book-as-a-commodity*: in effect, printed books have aesthetic and monetary value and, as such, other issues are raised, such as the question of originality, authorship and copyright. New notions came about in the wake of printing, such as the edition of books (e.g., first, second edition), their physical and material properties (e.g., bindings), with a significant increase in impact on society as a whole, for example the goal of global literacy.

Speech and writing are different, as they perform mode-specific functions and play specialized roles in society. In the subsequent paragraphs, I will examine their opposing features from a linguistic standpoint, singling out aspects for more fine-grained analyses developed in the following chapters.

1.3.1   Speech Is Natural, Writing Is Taught

Speech comes before writing from both a phylogenetic and an ontogenetic point of view. Ontogenetically, it is learnt naturally, as it does not need formal instruction (Halliday 1984a). Despite the fact that consensus is far from being reached with regard to the ways in which children learn how to speak a language (cf. behaviourist approach; cognitive approach, nativist approach, Chomsky 1957; interaction/functional approaches, Halliday 1978), it is undeniable that learning occurs in a varied set of informal contexts, first and foremost within the family: children learn with the help of their parents, relatives and caretakers. Conversely, writing requires effort to be learnt and such learning usually takes place in formal contexts, such as schools or similar institutions. As a consequence, speech is learnt more rapidly, as this process matures at a very early age, when brain development is very rapid, whereas learning writing happens at a later stage, when such a process has slowed down.

Furthermore, speech has evolved along with the development of human beings as a species, whereas writing is a technology that has emerged much more recently in the history of mankind. Learning writing means mastering a range of skills, such as producing graphic signs associated with sounds according to a shared code. These tasks are performed materially by hand or by typing and applying other related sub-skills, like adopting correct spelling, punctuation, capitalization and grammar (Cornbleet and Carter 2001).

Speech is far more widespread than writing, both from an individual (i.e., people usually speak more than they write) and a social standpoint, whereas literacy is commonly and methodically pursued only in some (industrial) countries, as discussed previously. Most notably, literacy is declining today even in the so-called "developed" countries and the issue related to education in a special form of literacy, namely digital literacy, is currently high on the research priorities of scholars in fields such as applied linguistics, multimodality, education and media studies (Kress 2003; McKenna and Richards 2003; Selber 2004; Lankshear and Knobel 2006).

To summarize, speech is natural and writing is taught. As a matter of fact, this simple observation may explain many of the social and cultural aspects of the oral/written debate as discussed in section 1.2. Following this observation, another difference emerges: speech is context-related, whereas writing is relatively context-free.

1.3.2   Speech Is Context-related, Writing Is Context-free

Consistent with the system through which it is learnt, that is open and not institutionalized, speech heavily relies on the context of production for interpretation. Linguistic features, such as deictics or discourse markers (Schiffrin 1987), refer to the context in which they are produced and speakers are able to interpret such features in and through a real context of situation: speech is immersed in a live context, it shapes it and is shaped by it. If I say "this" or "there," only the linguistic and situational surroundings can explain what I mean or to whom and what I am referring. Hence, speech and context conflate in the meaning-making event, as speakers are usually embedded in the same time and place in spontaneous and synchronous interaction. Conversely, writing operates at a *distance*, as writers and readers are typically separated by place and time (Ong 1967, 1977; Coulmas 1989).

Furthermore, context may be defined in both linguistic (endophoric) and extra-linguistic (exophoric) terms. A linguistic context, also called cotext (Yule 2010), is what surrounds each utterance, or, to put the matter in simple terms, what comes before and what comes after any given chunk of either spoken or written language. A linguistic context allows the interpretation of "He's awesome today," permitting, at the same time, the understanding of who is the unnamed "he" *and* possible reasons why "he is so awesome" in the speaker's opinion (dress, behaviour, facial expressions, etc.). Furthermore, as a speech event is commonly carried out face-to-face, it is also possible to convey extra meanings through non-verbal language, that is using gestures, posture, gaze or even body proximity, or by the conscious or unconscious use of social distance (i.e., proxemics). These non-verbal linguistic options, which are fully communicative in face-to-face interactions, will be taken up in Chapter 2, as they are especially relevant in online synchronous communication.

An extra-linguistic or local-situational context is made up of several extra-linguistic components affecting interaction, such as time, place, participant's role, gender, age and their relative status (Hewings and Hewings 2005). Of course, not all these elements are always relevant to every interaction. Kinds of activity influence the language used (cf. the concept of "field" in Halliday 1978) and all linguistic features, be they phonological, prosodic, lexical, syntactic or semantic, serve as contextualization cues. The latter signal which aspect is relevant to the context (Gumpertz 1982). In other words, context, in broad terms, including linguistic, situational, psychological and social components, shapes the language used, and, in turn, language affects the context in which it is produced.

Topics, register, relationships among speakers, and their relative status and purposes for interaction will also depend on more far-reaching notions, such as cultural expectations, for example those derived by the roles filled by participants and values shared in a speech community, determining what is allowed and what is not in any speech event. However, even though participants may be not equal in status, most forms of oral discourse share basic tenets, such as the possibility to ask and answer a question, to clarify and expand a point of discussion, to take turns and swap roles, and so on. The idea that these characteristics are shared and somewhat mandatory in speech events underpins several linguistic approaches to speech, such as conversation analysis, discourse and genre analysis.

Writing, in turn, raises a different concern with context, and it would be inappropriate (Cazden 1989) to label writing as completely context-free. The idea that writing is autonomous from context has been claimed in some quarters, as discussed in section 1.2; this idea is grounded, among others, in a long-standing tradition that considers writing as *absence,* as opposed to speech as *presence* (Zumthor 1984, [1983] 1990).

1.3.3 Speech Stands for Presence, Writing Stands for Absence

Participants are physically present in a face-to-face conversation, whereas writers and their readers do not share a physical co-presence. Another significant contextual factor is the distinctive presence of an *audience* in both spoken and written interaction. Recalling what I have said as regards audience and performance in oral societies, we may assume that those who will listen to, and participate in, what we say, affect how and what we say, in other words, the message and how it is delivered.

In general, participants are in some way acquainted with each other in spontaneous, face-to-face encounters, but writers may have little or no prior knowledge of their readership. Writers, however, usually do hold various kinds of pre-assumptions about their readers, based on certain shared values, cultural backgrounds and expectations about the genre, or text type. For example, journalists may have certain expectations about their readers, grounded on their political affiliations or cultural preferences. Novelists may also have their ideal readers in mind, even though the *ideal* reader does not necessarily correspond to the *real* reader.

However, contextualization cues are needed more in writing than in speech, the latter being, as I have said, more dynamically linked to a live context. Conversely, writers need to fill all potential gaps for the readers' benefit and ease the process of reading by gauging and lowering potential textual barriers. Cooperation is in both modes indispensable, but it is shaped differently according to situation types and other contextual variables. Furthermore, the nature of the relationship between speakers and writers/readers is completely different. A speaker has the opportunity to adapt actively to the other's assumptions, expectations and presuppositions, whereas writers are, more often than not, in the dark as regards their future reception. Immediate versus differed reciprocity ineludibly permeates spoken and written discourse.

Cooperation is a process of continuous and flexible adaptation to the other partner in live interaction, and participants need to comply with common and shared rules (Grice 1952); the same may be said about what readers do, with the caveat that they do not receive immediate feedback and are left alone with their interpretations, which the writer cannot either contest or approve. A consequence of this state of affairs is that writing is more detached and distant than is speech.

1.3.4   Speech Is Involving, Writing Is Detaching

As discussed in section 1.3.3, writing is generally held to be a solitary activity, at least in the context of traditional notions of writing. Face-to-face interactions may in effect have different purposes, one of which is *phatic*, that is helping to establish and maintain social relationships or carrying out a social task (Malinowski 1923). From this point of view, phatic communication or *small talk* (Laver 1975; Ventola 1979) is a typical example of *involvement* in interaction, as it deploys interactional communicative resources that are needed to keep up with social requirements (e.g., establishing one's and other's social roles, introducing and closing a conversation, filling silence or pauses in conversations).

Involvement, as epitomized in small talk, is thus contrasted with the detachment of writing, and has a twofold dimension: a) it is established with the listener, and b) also entails the relationship one establishes with oneself. Involvement, in other words, implies giving and receiving immediate feedback and also forms of self-reference that are usually avoided in writing (Biber 1988, 1995).

Writing, on the contrary, implies *detachment*, as a face-to-face rapport is excluded and readers find themselves engaged in the process of interpreting pre-fabricated fictional worlds, instead of coping with interpersonal relationships (Chafe 1982; Coulmas 1989).

However, things are not as straightforward as one may imagine with the qualities of involvement in speech and detachment in writing. For example, a lecture can be delivered impersonally, whereas an Internet Relay Chat (IRC) can be as immediate and interactive as a face-to-face conversation. Other traditional written genres are also influenced by speech, for example, in fiction. Tannen explains, "[C]reative writing is a genre which is necessarily written but which makes use of features associated with oral language because it depends for its effect on interpersonal involvement or the sense of identification between the writer or the characters and the reader" (Tannen 1982b: 14). In her view, speech and writing are intertwined; oral features, deemed as *involving strategies for interaction*, are used by writers to engross their audience in the story.

It is the *different level of involvement* in both spoken and written genres that makes the difference: the mode is not ultimately held responsible for degrees of involvement, but the strategies used to express higher or lower levels of involvement are (Tannen 1982a, 1989). However, research literature has also provided quantitative and qualitative evidence for preferred lexicogrammatical items in both modes, such as, to name but the most prominent, the use of personal pronouns and active voice in speech, and the use of nominalizations and passive voice in writing (Biber 1988).

Features of involvement and distance have been associated with different genres in both oral and literate cultures. For example, stigmata attached to alleged excessive involvement in oral narratives was instrumentally used to belittle nonliterate societies (Bernal 1987). It goes without saying that involvement matches *repetition*; that is one of the most distinguishing characteristics of speech and is considered unacceptable in most written genres.

1.3.5 Speech Is Redundant, Writing Is Compact

When we talk, especially in spontaneous conversations, we naturally repeat the same concepts over and over again, recycling the same wordings: our concern is to establish a rapport with our listeners, continuously checking their attention, approval, understanding, and so on. Repetition signals the unplanned, improvised nature of spoken interaction. Broadly speaking, the lexis of spoken discourse tends to be informal and is characterized by vague and redundant items, colloquial idioms, abbreviations and several recapitulation words that would sound odd and even unbearable in writing. Adherence to standard norms is also looser, and preferential grammar patterns include minor sentences and coordination, phrasal verbs and contracted forms. Conformity to standard varieties, which are usually and explicitly described as *written normativity*, is less expected in informal interactions.

The unplanned nature of speech is displayed in typical examples of spontaneous interaction. Of course, if we think of hybrid genres, such as a written informal note to a friend or a political speech, things run exactly counter to these descriptions. However, several studies attest the redundant and iterative nature of speech, as opposed to the concise and strict standardization of written discourse.

Unlike speech, writing is planned, editable and revisable. The process of revision, naturally, affects not only the formality and care applied, with consequent pressure put on the writer to be formally correct, but also the typical compactness expected from formal written discourse. Repetition usually enhances coherence and involvement in spoken interaction and especially so in oral societies (Tannen 1989; Finnegan [1978] 1992), whereas writing makes use of other linguistic devices for the same purposes, for example through anaphoric and cataphoric references (Biber 1988).

1.3.6   Speech Is Other-paced, Writing Is Self-paced

Another factor related to those discussed so far is the *difference in pace*. Writing, as well as reading, is *self-paced*. This means that one can plan, write, edit, revise at one's own pleasure, and as many times as deemed useful or necessary. But this is not the case with speech. A conversation is dictated by co-constructed time, pace and efforts. We take turns and swap roles, in a dynamic relationship that is instantiated by each participant. Speech, as such, is *other-paced*, as it implies a continuous negotiation of turns, and language usage varies accordingly.

1.3.7   Speech Is Evanescent, Writing Is Permanent

In the orality/writing theoretical debate discussed in section 1.2.1, the contrast between the volatile nature of speech as opposed to the permanence exhibited by writing is evident. Speech is evanescent and words dissolve as soon as they are uttered. But if we consider how technologies have made possible the reproduction of speech, then this opposition becomes meaningless. Video recorders and videotapes in the recent past, and a range of different digital devices today, have allowed reproducibility and thus the possibility to keep and store audio and video data (for an early discussion on the question of reproducibility, see Benjamin [1935] 2008).

However, if we take natural, spontaneous and face-to-face interaction into account, spoken discourse is neatly contrasted to writing in terms of permanence. Speech *can* be stored, but this is optional, whereas writing needs a material implement to exist. This has important implications for communication as a whole.

Writers can revise, edit, rewrite and so on, but as soon as the final *object* of writing is completed, further changes are virtually impossible. This is a generalization, but it cannot be overstated that, in traditional terms, writing is durable and permanent. Readers themselves are also involved in the process of storing and fixing written texts. Not only do they materially own and keep the physical object, the *book-as-a-commodity*, but they can go back, reread and recheck what they have read, and can even use a material implement to add their own *marginalia*, be it in the form of notes, glosses, comments, and so on. As I have suggested, interaction is practically absent with the book's author, but some degree of interaction with the text itself is indeed possible.

The same cannot be said for speech. Conversation can be neither "frozen" nor revised. Everything happens in real time and cannot be fully controlled and monitored. Speech fluctuates between adjustment and negotiation; once uttered, words cannot be unsaid or deleted. This echoes the opening of this chapter, namely the Latin proverb contrasting the idea that whereas speech is ephemeral, writing is permanent.

As has been argued, there are several ways in which these systems of production and reception affect both production and understanding, for example the idea that short-term memory is more important in speech than in writing (Nickerson 1981). Others have claimed that some structures are more likely to occur in writing than in speech, for example the use of nouns rather than verbs, due to the fact that writing objectifies the object-text by nominalization. Halliday (1985b, 1987), for example, claims that preferential structures for each mode may be ascribed to their different nature, because a written text is an object and, as such, tends to objectify and crystallize concepts and phenomena, also by changing the class of words through grammatical metaphor (Halliday 1984b, 1998).

This opposition between the permanence of writing and impermanence of speech is also important in relation to the different status each of them acquires in society. As I said beforehand, the notion itself of Standard English hinges on socially agreed upon written varieties. Summarizing research literature on the matter, Crystal (2003: 110) identifies some essential characteristics of Standard English, listing linguistic features such as lexis, grammar and orthography (i.e., spelling and punctuation), and explicitly putting aside speech-related criteria, such as pronunciation and accent. As has been widely argued (Hogg, Blake, and Algeo 2001; Crowley [1989] 2003; Wright 2006), the codification of a language is fundamentally linked to the development of its writing system. Descriptions of language themselves are mostly found in written forms, for example in grammar books, dictionaries, and so on, and nowadays electronic and digital devices are also extensively used to store language information and descriptions. Speech is thus more prone to change, as it resists the fixity implied in codification, whereas writing and codification are connected.

1.3.8   Materiality of Speech and Writing

What are the linguistic characteristics of each mode? In their abstract versions, several important differences emerge. The most evident is the distinction between mode-specific *physical* realizations. Speech is based on prosodic features, such as intonation, loudness, pitch, tempo, rhythm, stress and pauses. Through prosody, information of various kinds is conveyed, for example with regard to a speaker's feelings and attitudes. Furthermore, intonation displays a grammatical function (e.g., placement of boundaries between phrases and clauses) and a discourse function (e.g., the difference between "new" and "given" information) (Roach 2000).

In effect, non-verbal meaning is brought about not only by traits surrounding language, such as those listed beforehand, such as gaze, posture, and so on, but also by traits *within* language, that is, related to *how* we say something. In this light, prosody is fully meaningful. By varying intonation, for example, speakers can convey meanings by means of different grammatical structures, such as statements, questions, and so on: "She is late, right?" is a question, even though it is not syntactically structured as such. Intonation is also useful in communicating meanings related to the context in which they are produced, but is also an important marker of social and cultural identities. For example, some professionals, such as lawyers and sports commentators, are easily recognizable by their distinctive prosody. Crystal defines this function of prosody as "indexical" (Crystal 2003: 249).

Variations in pitch are usually linked to the involvement of the speaker in the subject, while variations in stress mark rhythm. Pace and loudness also impart meaning and attitude. Pauses and hesitations are also fully meaningful, as we are all well aware, and they give structure and organization to the otherwise fluid and evanescent speech.

All these elements cannot be always and systematically reproduced in writing. Writing systems have developed alternative strategies: for example, signalling a question with a question mark, or pauses and hesitations by periods or suspension points. Stress can be reproduced in writing with graphical devices like **bold** or underline. But the idea that writing should be downgraded if compared to speech on these grounds is untenable.

Writing thus displays different unique physical/material features. What is performed by prosody in speech is performed by graphical features in writing, which are currently in the process of unprecedented expansion in the digital world. Graphic symbols have been stretched today more than ever to express a wide variety of visual meanings. Such physical characteristics in writing are actually represented by visual signs that give shape and substance to words, for instance creating hierarchies (e.g., capitalization, bullet points) or, as in the case of digital communication, allowing the addition of a personal touch (e.g., use of colour, layout, emoticons). Writing thus possesses distinct attributes that help with parsing phrases and meanings.

Hence, each mode exhibits different ways of giving shape and structuring meanings. Strategies are not easily comparable. A question mark signals a question in writing, but there are many other ways that can be employed in speech to mark the same kind of grammatical structure. By the same token, writing can also affect the way we speak: for example, it is not infrequent to mimic quotation marks with the fingers to emphasize the speaker's attitude toward what is said (e.g., irony, sarcasm, doubt). In effect, prosody and punctuation perform several functions; each of them is mode-specific and, as such, is more suited to express a certain role or attitude (Chafe 1988).

Way of signalling different levels, or orders, of hierarchies are practically absent in speech: indentation or the use of distinctive fonts or layout are used in writing to create such organizations and categorizations that, according to some scholars, have facilitated the formation of sustained and analytic thought in the human kind, thanks to systematization of knowledge made possible by writing (Ong 1992).

But what are the preferred lexicogrammar and linguistic structures for each mode?

1.3.9   Preferential Lexicogrammar in Speech and Writing

In speech, there is no time-lag between production and reception. The speed of spontaneous face-to-face interaction does not allow for advance planning, as writing does. We think and talk almost at the same time and, as a consequence of this on-line process, looser constructions are preferred, including redundant and repetitive constructions and lexical items, rephrasing and comment clauses. Writers, conversely, have time to plan and revise, even though they usually do not have direct contact with their readers to check understanding. Careful planning is allowed, including the possibility to avoid repetition and construct more polished, varied and intricate structures.

Research literature has found that written texts in English include longer words, a more varied lexis (Drieman 1962; Blankenship 1962, 1974; Devito 1967a, 1967b; Olson 1977; Chafe and Danielewicz 1987; Hayes 1988; Tottie 1991; Stein 1992; Yates 1996), more content or lexical words (Halliday 1985b, 1987; Stubbs 1996), more complex noun clauses and nominalization (Biber 1988; J. Miller and Weinert 1998) and more subordination (Hesch 1982; Altenberg 1984; Beaman 1984; J. Miller 1993; Smith 1994). Furthermore, written texts usually display more adverbial and prepositional phrases, more complement and relative clauses, more indirect questions and quotations (Chafe 1982, 1983, 1985; Chafe and Danielewicz 1987). Written language is, as a consequence, more precise and exact than is speech, also because the durability of writing puts pressure on writers to produce valid, accountable and accurate statements.

Speech, by contrast, is less planned than writing, and speakers do not have full awareness of their discourse before they have uttered it. They are also prone to errors, disfluencies, fillers, false starts, hesitations, stuttering, repairs, and also a number of fuzzy/vague words and expressions, such as "kind of," "sort of," "like" (Chafe 1982; Biber 1988; Smith 1994) and vague pronouns, such as "it" (Biber 1988, 1992).

Many attempts have been made to turn the speech-writing debate into a more exploratory and explanatory labelling cutting across the two modes, for example in the opposition between integration vs. involvement (Chafe 1982, 1983), self-monitored vs. spontaneous (Halliday 1985b, 1987), focused vs. non focused (Scollon and Scollon 1984), contextualized vs. decontextualized (Denny 1991), planned vs. unplanned (Ochs 1979) and formal vs. informal (Akinnaso 1985), all of them made significant *across modes*.

Biber carried out multivariate factor analysis using methods from corpus linguistics to analyse the major dimensions of speech and writing, concluding that *stereotypical speech* and *stereotypical writing* can be postulated on the grounds of variation across dimensions, found in different text genres. He contended that variation could be measured and accounted for within the English language and also across other languages (Biber 1988, 1995). Biber's original framework (1988) involves the study of sixty-seven linguistic features, all functionally related. As he explains: "Methodologically, the approach uses computer-based text corpora, computational tools to identify linguistic features in texts, and multivariate statistical techniques to analyse the co-occurrence relations among linguistic features, thereby identifying underlying dimensions of variation in a language" (Biber 1995: 314). He includes a wide range of linguistic features, as no single linguistic parameter is sufficient in itself to grasp the variety of similarities and differences in the spoken-written *continuum*.

With his thorough review and discussion of research literature on the matter, Jahandarie (1999) argues that linguistic evidence indicates that separate sets of attributes can be established for each mode. In other words, the distinction made between speech and writing is there and there is no way of getting rid of it, despite the efforts to reduce dichotomies. Most attributes related to speech refer to "interactiveness, evanescence, 'on-the-fly' production and the use of prosody" (1999: 149), whereas writing exhibits "solitary, permanent and planned" qualities (1999: 149). He concludes that "even though it has proved impossible to find a precise demarcation that would separate all spoken from written genres [. . .], there are very clear patterns of association between each modality and different linguistic structures that point to their relative independence" (1999: 149).

This independence forms a part of the theoretical ground for this research. I have started by addressing the question as to whether the opposition in language modes is meaningful. Reviewing research literature, I have drawn a line of distinction between speech and writing, when dealing with *stereotypical*, to borrow from Biber (1988), versions of both modes.

Table 1.1 Mode-specific features in speech and writing

Speech	Writing
Concerned with *here-and-now*.	Concerned with *there-and-hereafter*.
Learnt naturally, in a familiar/informal context. In normal conditions, it is usually learnt at a very early age.	Learnt at school or similar institution, in a formal context. Usually implies formal instruction.
Heavily affects and is affected by the surrounding context. Participants are usually present and context is co-shared. Speakers know, or have some kind of contextual contact with, their listeners. Situated.	Relatively context-free. Participants are usually in different places. Writers may have ideal readers in mind, but they never know who will be the real/final addressee. Atemporal and asituated.
Focused on involving participants. Engaged in casual, unplanned and phatic communication. Helps create and maintain social relationships. Also relies on non-verbal language (e.g., posture, gaze, body movements).	Focused on detaching the subject-matter. Creates distance so as to provide information, discusses facts and records experience to the benefit of larger audience/s. Also relies on visual language, such as tables, charts, diagrams.
Repetitive, cyclic, dynamic. Prefers redundant structures, such as coordinate clauses.	Compact, concise, static. Prefers compacting functions of language.
Other-paced. Time-lag is minimal, so participants' responses are constrained by other participants' prompts.	Self-paced. Time-lag is decided by both writers and readers, who can, respectively, produce and receive written texts at their own pace and preference.
Impermanent and volatile. More subject to change, both in extemporaneous conversation and in terms of diachronic language change.	Permanent and durable. Less subject to change due to the materiality of implements (e.g., stone, paper, electronic) and thanks to the possibility of codification (cf. notions of standardization and Standard).
Relies on sound and voice prosody (e.g., intonation, pitch, rhythm, tempo) and may present numerous disfluencies due to its unplanned nature (e.g., false starts, hesitations, pauses, repairs).	Relies on a shared graphic system of representation (e.g., alphabetic, syllabic or logographic scripts, punctuation, spelling, subdivision into smaller units, e.g., paragraphs). Highly codified and less subject to change, unless academies or language planners dictate reforms in writing (e.g., spelling reforms).

However, the implication from section 1.2 is that no mode should be deemed as intrinsically "superior" to the other. Of course, speech came first, from an ontogenetic and phylogenetic point of view. But none of them holds primacy over the other. Both prestige and stigmatization come from social, political and ideological constructions. There is nothing intrinsically "superior" from a linguistic point of view. Even though the idea of a cline, or *continuum*, to explore and analyse speech and writing is equally useful, it is nonetheless true that both modes possess what I have called *mode-specific features*. Linguistic evidence, as discussed in this section, further supports this hypothesis. Table 1.1 provides a concise summary of the properties discussed in this section.

In the following section, all these notions will be applied to the domain of discourse of digital environments. By changing the medium, we will see which modification affects which mode.

1.4   SPEECH AND WRITING IN THE DIGITAL AGE

How have modes been modified by the advent of new technologies? Martinec and van Leeuwen (2009: 1) argue that the new media differ from language in three ways: (1) they are multimodal, because they are made up of different semiotic resources; (2) they are non-linear, because they combine spatial and temporal patterns; and (3) they are new, because they diverge from what used to be taken for granted regarding language.

Starting from the last consideration, which seems somewhat vague, I will revise the notions presented in the previous section to see if and how they need to be adjusted to such new digital environments. At the Georgetown University Round Table on Language and Linguistics (2011), Herring argued that research on communication on digital environments often claims that dealing with "new" environments calls for new theories, methods and systems of analysis (Herring 2013). Are the media discussed and presented in this volume *new*, and if so, to what extent and in relation to what?

Introducing video relay chats, blogging, social networking websites and media sharing communities, a common denominator seems to be the domain of Web 2.0, which needs to be addressed to enquire into the *new* media and the epistemological necessity to call for new descriptive models and theories. A generally agreed upon definition of Web 2.0 emphasizes ideas such as participatory information sharing and collaboration, user-generated content, and the web as a platform. However, these ideas are controversial, as are claims as to which digital environments are *new*.

Digital communities have been investigated by a number of studies, but it is beyond the scope of this work to compile a full review of research literature on the matter. Among the most significant studies, however, we may mention Thurlow and Mroczec (2011), who gather research on digital media, multimodal studies and linguistics, as well as Tannen and Trester

(2013), about "discourse 2.0." Jewitt's handbook on multimodal studies is also a useful collection on cutting-edge research (2009), whereas O'Halloran (2004) and Scollon and LeVine (2004) introduce multimodal discourse analysis. O'Halloran and Smith (2011), who collect contributions on multimodal practices within the domain of interactive digital media, provide significant and recent contributions. Reflections on how oral language has been misunderstood in debates on digital media are given in Gee and Hayes (2011). They claim at the outset, "digital media are an interesting hybrid of the properties of oral language and of written language" (2011: 1). They also argue that digital media have powered up language, increasing its intrinsic communicative power, and bestowing new abilities at the same time. However, technological affordances, in their view, also involve losses. Kress (2010) is concerned with gains and losses brought about by new systems of communication. Technologies are, in his view, defined as cultural resources, "[T]hey are taken up or not; inserted or not into life-worlds by social agents according to actual or felt social requirements and constraints. In that, they follow and foster contemporary social transformations while being shaped by them" (Kress 2010: 195).

Several studies address the questions of education and technologies, in the light of new focus on literacy studies, which awaken interest in, and explore, new forms and notions of literacy in a globally digitalizing world (Selber 2004; Carrington and Robinson 2009; Gee and Hayes 2011; Selwyn 2011). Learning how to speak and write in the digital world is rather different from what happens in the physical world. Different skills are required, even though major similarities can be found in digital forms of speech and writing, if compared to their traditional counterparts. At GURT 2011, Baron enumerated some of the contemporary challenges to print cultures, namely related to *durability* (e.g., cheap newsprint, paperbacks), to the *conditions of reading* (e.g., multitasking), to the value of books (e.g., speed reading, libraries removing books and journals and replacing them with e-books and e-journals, issues of copyright), to navigation tools (e.g., apotheosis of random access, reduction in page numbers).

All these questions are connected to education and literacy. What is the future of teaching writing? According to Baron, teaching writing has changed dramatically in the past few years, as the shift to communicative approaches in foreign language learning has tended to put more emphasis on writing, aiming at self-expression, thus neglecting the rules of "good writing" (2010).

So what has happened to literacy in this scenario? In debating the thorny question of literacy, Kress wonders "what direction writing is likely to move: will it move back towards speech-like forms, and become mere transcription of speech again, or will it move back in the direction of its image origins?" (2003: 61). His argument centres around the idea that language-as-speech will be the most used means of communication, whereas language-as-writing will be displaced by the prominence of the image, which, in his view, is more useful in defining communication today. In effect, from what I have discussed so far, language appears to be investigated predominantly through its *verbal* apparatus and vestiges, whereas our everyday experience tells us that communication is much more than this.

Speech and writing, seen from this point of view, provide a "shell" for verbal language, leaving aside all other aspects that are equally significant in communication, as multimodal studies attest. Much in line with previous bodies of research, Kress discusses the distinct properties of speech, writing and image, calling them *logics* and *affordances*. He makes the point that in earlier times, the association mode-medium was immediate and automatic: writing (mode) was easily and univocally associated with a book. This is particularly interesting in this context, as different communities or groups of participants in digital environments are sometimes lost in the apparently indiscernible richness of modes. Such associations of mode-medium have become completely meaningless.

Whereas the logic of both speech and writing rely on spatial/sequential elements, the logic of image (and screen) is spatial/simultaneous. In writing, the meaning comes out from the sequential arrangement of verbal items as they unfold in time, but in image, the meaning derives from the spatial relationships of the depicted elements.

1.4.1 Materiality of Digital Speech and Writing

The crucial disentanglement between mode and medium thus results in a much more complex framework, where many of the previous well-established notions no longer hold. Notice, for example, how the semiotic affordances of image and screen alter what I disputed regarding speech and writing. Furthermore, it is important to be fully aware of the *material qualities* of speech and writing. Despite the fact that many differences appear as completely diluted in the web scenario, material qualities, such as prosody countervailing script, for instance, still hold. Speech is persistently produced by human voice (in human communication) and writing continues to make use of a script, made up of graphic symbols, to produce meanings. This fact runs the risk of being too banal, not worth mentioning, but I believe that such materiality also shapes web-based interaction in significant ways. The simple observation of such irreducible difference is well worth taking into account in any analysis of spoken and written discourse in digital environments.

As for the difference in materiality, voice and sound are opposed to a graphic technology based on a script in speech and writing, whereas a digital text is made up of bits and bytes, a data technology that makes use of discrete values, that can be either discontinuous (e.g., numbers, letters, symbols) or continuous (e.g., images, sounds, etc.). Digital technologies are contrasted to analogue (or non-digital) technologies and digital texts embed both speech and writing in their systems of producing meaning and construing interactional social practices. The relationship with context is, as I have said, negligible or null in digital environments, because it is sometimes impossible to reconstruct all the bits and pieces of an interaction, which is spatially and temporally fragmented (e.g., participants are located in different places and, in the case of asynchronous interaction, at different times), and also including partial or null contextual cues—it is possible to view *parts* of our interlocutor's context, which is virtually projected on screen. In her book on alphabet and email (2000), Baron quotes an illuminating passage from Isaac Asimov's short story "Solaris," where a character points out the difference between seeing interlocutors face-to-face and viewing them onscreen. When another character retorts that *seeing* and *viewing* is exactly the same, she replies that he was not seeing her, that would have been possible only face-to-face, he was just *viewing her image*. My point is that this difference is linked to materiality and the possibility versus impossibility to share the same context.

Let us pause for a moment and revise the notions of speech and writing in the digital arena. One of the most important differences lies in the different localization of *context*.

1.4.2   Context, Localization and Authorship

Context in the new media is far from easy to grasp and define. The notion itself loses ground as digital communities are located virtually anywhere and nowhere at the same time. All the material features making up the context, where face-to-face interaction takes place, are absent. A physical location co-shared by a group of participants is replaced by an anonymous *nowhere*, made up of bits and bytes. In this sense, the abstract concept of community has also been challenged (Herring 2004), given the amorphous membership, low social accountability and lack of co-shared geographical space (McLaughlin, Osborne, and Smith 1995).

In this light, context is a fluid notion on the internet and, therefore, speech and writing tend to lose their own links with context. Speech may be recorded and reproduced, but *it is also typically uttered in one context and received in another*. A relay chat or a video produced by a user and posted on a video sharing community or a social network is partly made up of speech, but its reception does not rely on a shared context of situation. Such a spoken text is attended by an intended participant, an unintended group of participants or a combination of both, much as it is with book production. From this standpoint, speech becomes more like writing in its lack of direct contact with an unmediated context.

Connected to these considerations is the notion of localization, which needs to be distinguished from context, as context is a richer notion, including space, time, and other cues. Localization, conversely, is a narrower notion, dealing with the place where communication takes place. Whereas localization is high in speech, it is low in writing, and in digital texts, it may be low or null. This means that when interaction is face-to-face, participants share the same physical setting. This is not the case in common written discourse, but it is nonetheless possible to be cognizant of the addressee's localization. In digital texts, this is still possible, but much less likely.

Directly linked to localization in speech is the question of authorship. As producer and receiver are co-present during a spontaneous conversation/ interaction, the identification between producer and utterance is immediate and unambiguous. In writing, the idea of authorship is linked to the iden- tification and equation of the written work with its author/s. This notion is relatively recent and was exalted in the Romantic period, when the semi- nal notions of originality and authorship were so important. Authorship in digital texts achieves both the meanings of association between author and utterance and accountability: we may or may not know the identity of the author of a text (depending on the presence of a webcam in synchronous interaction or a video clip in asynchronous communication), but even if we *view* our interlocutors, we may still not know who they are. Furthermore, the text may be autographed or anonymous, thus making the association between author and text impossible.

The question of permanence has been challenged to a great extent in the digital age. If a distinction between the volatile nature of speech against the permanence of writing held for most of the twentieth century, today the aura of permanence of writing seems to have lost some of its appeal. Permanence in effect involves the durability of the implement/material (e.g., stone, parchment, paper) and the inalterability of written texts, protected by the material support and by other socially constrained factors, such as copyright and authorship.

1.4.3   Involvement and Informality

Another complete modification of roles concerns the described dichotomy between involving strategies of speech and detaching strategies of writing. In this case, writing surprisingly comes closer to speech in web-based texts. Online interactions are usually much more friendly, open and informal, and communication usually endorses the production of rapid, unrevised and dy- namic texts (both spoken and written) that, in one way or another, establish (or apparently do so) virtual social bonds of several kinds. Involving strate- gies are pursued to create and sustain such bonds. The detached style used in formal writing gives way to interactive strategies used more typically in conversation.

Directly connected to this feature is the preferred lexicogrammar used to communicate online: speech-like features, such as the use of an unplanned and unrevised language, are used in digital informal communication. Writ- ten language has in effect undergone a profound change on the internet. Influential studies, such as those carried out by Baron (2000, 2010), have shown that contemporary writing is increasingly informal: in her view, this informality derives from the tendency to use writing in a way that approximates speech (Baron 2010).

Even though in many cases computer-mediated communication is written, it is usually more associated with speech, as they are both ephemeral and impermanent. However, digital writing is so easily produced and deleted that it is easily associated with speech.

According to Danet (2010: 146), if compared to speech and conventional (i.e., pre-digital) writing, digital writing is both *doubly attenuated* and *doubly enhanced*. It is attenuated because it lacks materiality (and durability), but is nonetheless enhanced, as it leaves traces, which speech does not. Danet lists some typical features of computer-mediated English language, adapted and enriched in Table 1.2, with examples taken from the corpora used in this book and with the addition of descriptive functional categories.

Many studies have in effect highlighted the jocularity typical of computer-mediated communication. The aforementioned informality commonly

Table 1.2   Features typical of computer-mediated English language

Feature	Example	Function
Reduplicate punctuation marks	What????!!!!?????	Emphasis, playfulness
Reduplicate phonemes	Whaaaat? Aaaaaaaaaaarghhhhhh	Emphasis, playfulness
Non-standard spelling	Txt me, 4U, bizy	Emphasis, playfulness
Capitalization	DO NOT SHOUT AT ME PLEASE FOLKS did u miss me?	Expression of anger, resentment, aggressive behaviour or, if it concerns a single word, emphasis
Lowercase	i need more time as you know. well, this sounds	Speed writing, sloppiness
Acronyms, shortenings, abbreviations	LOL (laughing out loud); bff (best friend forever)	Playfulness, speed writing
Descriptions of actions, events, states of mind	"laughing"; <laughing>; *laughing*	Emphasis to a certain word or attitude
Emoticons and other visual items (e.g., sending a photo, a picture, a text file)	☺ ☹	Visual contents
Onomatopoeias	argh, ahahahah	Playfulness, mimics speech

Adapted from Danet (2010: 148).

found in speech is in this sense transferred in digital writing that lacks the typical formality of traditional writing and is enhanced, to borrow from Danet (2010), by imitating speech.

As will be discussed in the following chapter, playfulness and jocularity cannot be analysed in strictly linguistic terms. From one standpoint, we may argue that it stands in opposition to formality, but this does not tell the whole story, that is best described in multimodal terms, for example highlighting the visual nature of such digital modifications in writing, and the added value provided by visual elements (Sindoni 2011b). The *mimicking-speech* nature of digital writing is not exclusively achieved through the use of emoticons, which are extensively employed in online interactions to signal facial expressions, such as smiling, winking, frowning and nodding, but also by other signals of non-verbal behaviour, such as the use of exaggerated punctuation, which, in turn, mimics prosody. For example, variation in pitch is indicated by the use of capital letters or question marks, signposting intonational contours, such as in the expression of bewilderment or surprise.

The use of informal language, at least in the case of web-based interactions, such as those found on personal blogs, videochats and social networking communities, parallels the preference for unrevised and rapid language, using speech-like features, such as repetitive lexical items and coordination, instead of more refined and polished forms of traditional and formal written use, which is characterized by compact and concise style, as detailed in section 1.3.9.

1.4.4   Production, Reception and Interaction

The production of a spoken text is usually extemporaneous, as is its reception. In written texts, the production is, conversely, planned, and reception is delayed. In digital texts, everything depends on the degree of synchronicity of communication. If it is synchronous, production and reception are extemporaneous (even though perfect synchronicity is impossible on the web); if it is asynchronous, production is planned and reception is delayed, even though time for planning and responding are considerably shorter than in conventional written texts.

With regard to the time of production and reception of writing in digital environments, it is a blend of self-paced and other-paced multimodal constructions. When a synchronous interaction takes place, as in the case of Internet Relay Chat (IRC), the pace is other-oriented. In other words, texts are sent and received with a minimum time lag in IRC, as it is very similar to a face-to-face conversation, with extemporaneous and usually short exchanges, because participants' responses are constrained by the other participants' prompts. In this sense, web-based synchronous interaction is other-paced, as it may be associated with face-to-face interactions, because it is constrained by the interlocutor's prompts. Informants in this study have in more than one occasion reported that time lag in web-based synchronous interaction may be looser than in face-to-face conversations. It may happen that a response is not given in real time, as it would occur in a live interaction. However, it is considered acceptable by most users because they are well aware that many variables (e.g., technical issues, such as being offline for some time) affect conversation. From this standpoint, synchronous interactions are also partly self-paced, as writers and readers can also adjust their responses to their own preferences or needs.

For these reasons, the degree of interaction is high or very high in spontaneous conversation, whereas written communication tends to weaken the level of interaction, which is nonetheless retrieved in digital interaction that, in both spoken and written modes, includes high or very high levels of interaction both among participants and between participants and text (e.g., interactive texts, hypertexts, blogs, social networking communities). Interaction parallels the *control* that participants are able to exercise over their spoken, written or digital texts.

1.4.5 Degree of Control and Storage

Degree of control depends on several factors, such as the speed of producing and processing language that affects the power we have to produce fully articulated, appropriate and correct sentences or, conversely, the chance to make mistakes, use pauses, fillers and hesitations, in short being less able to check the online verbal process while it is produced. Of course, speech is more fluid and is produced and received in real time: on the one hand, speakers are less able to control it and, in spite of their use of corrective strategies, for example to fill embarrassing silences, their capacity to produce fully articulated and faultless sentences is less likely than educated people like to think. On the other hand, writing allows revisions, multiple readings and versions, so the degree of control, in the production process, is very high.

On the level of reception, however, control is almost nil in traditional written and published texts: writers can only publish a second edition of their book or, at best, compile in haste an *errata corrige*, otherwise their work remains unchanged with all its merits and faults . . . until the next edition.

Digital texts are an amalgam of these features: control is usually attenuated if compared to writing, as production is usually imbued with impatience and speed of composition, but it can in theory be more controlled than in speech, as revisions are possible and allowed. Digital texts may be also partly faithful to their spoken or written counterpart, but that is a question of constitutive differences.

As we have seen from the account of how oral and literate societies developed in section 1.2, the possibility of storing texts is fundamental not only for the existence and durability of texts, but also for attached aesthetic and moral values. Storage is naturally not expected in spontaneous spoken interaction: it is not impossible, as we can record or video-record a conversation, for example for research purposes; it is simply insignificant. Storage is not necessary in conversation or any face-to-face interaction, whereas the opposite happens with written discourse, which dictates the presence of a material support specifically linked to its durability.

The concept of storage becomes relevant in digital discourse, as web-based texts can be stored, for example, by recording a video conversation or downloading emails to an inbox, but this is optional. Paradoxically, with digital technologies, the capacity of storage is much greater than before: an entire library, or more than one library, can be stored *digitally* on a mobile device or on a laptop. However, much online interaction is volatile and, despite the possibility to store it, few take the trouble to do so, unless specific circumstances so require.

1.4.6   Provenance

The question of a link between a text and its producer/author introduces the concept of provenance. Where are the producers and where are the texts they have produced? Are they bound together or do they live independent lives?

In common everyday interaction, speakers are inextricably bound to the utterances they produce. Listeners are aware from whom the message originates. In the same way, readers are also aware of the author's identity and the book/article/letter's provenance in written discourse. However, once written texts are produced, they live an independent life from their authors, who lose their hold on their creations. Readers may read, comment, scribble or simply squander such texts.

In digital environments, the concept of provenance is more complex. Provenance may be known and not known, sometimes even at the same time, as readers may be aware of some contextual details, such as when and where the piece of writing was created and who the author is or where s/he comes from, even though such details are less monitored than in traditional print culture. Provenance has been distinguished from other categories, like authorship (only concerned with the author's identity and with claims of authorship on the work in the case of written literature) or context, which is a broader notion. Provenance is, on the other hand, more concerned with where the physical and symbolic sources of a spoken, written or digital text come from.

1.4.7   Use and Type of Multimodal Resources

The last categories we will discuss in this section deal with the use and kind of multimodal semiotic resources used. Notice that Kress and van Leeuwen

(1996, 2001, 2002) use the term *semiotic mode*, while O'Halloran (2004), O'Toole (2010) and Baldry and Thibault (2006) favour the term *semiotic resource*. The latter usage is adopted here, because *mode* is here intended in Hallidayan terms, as the mode used in language. Whereas some theorists, such as those quoted above, apply the concept of mode to all systems of meaning making, I prefer here to call these systems semiotic resources, with the specific goal of *differentiating* verbal language from other meaning-making resources (Sindoni 2011b).

Semiotic resources are used in meaning-making events to convey different messages and situation types endorse the use of specific resources appropriate to them. For example, in a spontaneous conversation, the distance we automatically put between ourselves and our interlocutor is a communicative strategy that conveys meanings, for example related to the degree of intimacy or social distance, as has been termed (Hall 1966). In this case, distance is not determined by chance, but is socially constructed and co-negotiated in a communicative event, or, to put it in simple terms, is a semiotic resource that we use to communicate meanings in a given circumstance to someone we interact with.

In another situation type, we may wish to use other resources to communicate to good effect. While writing a formal letter, we usually make use of layout, spacing, font and other graphic devices to express certain meanings, such as the degree of formality and our ability to master and conform to written standards, directly linked to our education and context of culture. No one would, for example, use a multicoloured font in a CV, unless an eccentric web or fashion designer was trying to make an impression on a job interviewer. Neither would we write an elaborate note to our addressees while we speak to them. However, things are less clear-cut in digital environments, as everyday web-based practices show. Participants can talk using writing as the main resource or publishing posts and mixing resources, such as videos, pictures, photos, and so on.

To what extent they use all these resources in the same communicative event, be it a videochat or a blog entry, and to what extent they are likely to integrate all these resources will be one of the highest priorities on the book's agenda.

Multimodal resources are thus here considered in their broadest sense: as meaning-making systems to convey meanings in all situation types. The degree of use of multimodal resources is thus high in spoken interactions, which usually imply the regular use of voice prosody, gestures, body positions and movements (i.e., kinesics), social distance (i.e., proxemics), and so on. In typical written interactions, the use of multimodal resources is maybe less obvious where conventional densely written and print works are concerned. However, such resources are nonetheless present in the use of layout, font, hierarchization of texts in visual units and subunits (e.g., body of text, paragraphs, etc.), and, maybe more evidently, in the placement of visual materials (Sindoni 2012).

Table 1.3   Mode-specific features for spoken, written and digital texts

Spoken text	Written text	Digital text
Materiality: voice and sound prosody	*Materiality*: graphic technology based on a script	*Materiality*: data technology based on discrete values
Context: high	*Context*: medium-low	*Context*: low
Localization: high	*Localization*: low	*Localization*: low to nil
Authorship: high. Speaker and spoken text inextricably bound	*Authorship*: high in contemporary texts (cf. copyright)	*Authorship*: high-medium-low. High in spoken interaction, medium-to-low in written informal interaction (cf. avatars, nicknames)
Degree of involvement: high	*Degree of involvement*: low	*Degree of involvement*: high
Production: extemporaneous	*Production*: planned	*Production*: extemporaneous and planned
Reception: extemporaneous	*Reception*: delayed	*Reception*: extemporaneous in synchronous interaction, delayed in asynchronous interaction
Degree of interaction: high	*Degree of interaction*: low	*Degree of interaction*: high
Degree of control: low	*Degree of control*: high	*Degree of control*: low-medium
Storage: nil in normal conditions, but possible with the aid of external supports (e.g., camcorder)	*Storage*: traditionally in material support (e.g., books)	*Storage*: in digital support, but online interaction is, more often than not, not stored
Provenance: known	*Provenance*: known	*Provenance*: known and unknown
Use of multimodal resources: high	*Use of multimodal resources*: medium-to-high	*Use of multimodal resources*: very high
Typical kind of multimodal resources: voice, gesture, body movements, distance, gaze, prosody, etc.	*Typical kind of multimodal resources*: page-constrained spatial arrangement of items, layout, use of written resources (e.g., font, colour, size), image	*Typical kind of multimodal resources*: screen-constrained spatial arrangements of items, voice mediated by the machine, "de-structured" body, layout, colour, image, written resources

In spite of this, a high degree of use of multimodal resources is without doubt employed in digital texts, where all these can be mixed, decoupled and reassembled in new or unprecedented configurations, thus fabricating creative interpersonal meanings. To name but one prominent example, which will be discussed in detail in the following chapter, think of the videochat, which allows participants to mix speech and writing at the same time. For this book's concerns, it is interesting to note that this possibility is exceptional in the history of human communication, as it was unthinkable to use both modes at the same time and in the same communicative event. Think also of blogs or media sharing communities, where users can play with semiotic resources they have at their disposal to produce unique, and often idiosyncratic texts.

Finally, the *typical kind* of multimodal resource used in spoken, written and digital texts are discussed. As has been anticipated, almost any semiotic resource can be used in spoken and written communicative events, and especially so in the digital scenario. However, we may resort to an exploratory categorization to make hypotheses on possible recursive linguistic, semiotic and multimodal patterns in each situation type and related texts. Predictions made possible by common sense inform us of the probability of finding voice, proxemic and kinetic patterns in spoken texts, and page-constrained spatial arrangements of items in written texts.

In web-based texts, item arrangements are screen-constrained and participants may be either hidden from their interlocutors' gaze or partially visible, thus deconstructing their bodies and face (or anything that is placed within the user's screen) becomes a metonymy for their whole bodies. Spoken and written resources are amply used, along with other visual systems (Sindoni 2010, 2011a).

Table 1.3 summarizes the aspects treated in this section, putting together spoken, written and digital texts. All of them are considered in their idealized versions, but it cannot be overstated that all these texts are far too complex and varied to be squeezed into such a tripartite categorization. However, this heuristic model is useful to approach these domains, but without overriding inescapable overlapping and juxtapositions.

1.5   DIGITAL EXPLORATIONS: AN OVERVIEW OF METHODS AND DATA

> Normally, whenever two people are in earshot, they speak to each other. Only very special circumstances—wicked children passing secret messages in class; partners who are "not talking" to each other; a jury foreman passing a verdict to a court official; someone who cannot speak or hear (and who is unable to use sign language)—would motivate the enormous trouble of writing down what we wish to "say." (Crystal 2003: 291)

As I have been discussing in this chapter, speech and writing have been studied using a wealth of methods and with several underlying reference frameworks of analysis. In the context of literate or nonliterate societies, the use of oral or written modes for aesthetic expression has never been casual, and the line of demarcation between literacy and illiteracy is blurred and often poisoned by questions of power, clashes among cultures and histories of cultural plunder perpetrated by the so-called Western societies. Speech and writing can in this sense become a symbol of the economic, socio-political and cultural divide between the two opposing facets of the globe. Literacy has been depicted in mainstream literature as a goal of civilization, and indeed it is, but gains of literacy do not come without losses.

In the present study, several methods will be applied, following the very general rationale according to which each specific instance of communication, be it spoken or written, may need different methods of enquiry that could save text-specific linguistic, communicative and, more broadly, interactional features. This book explores some digital web-based texts, and specifically multiparty videochat, blogs and social media sharing community practices.

These domains have been selected because they present *typical* or *stereotypical* forms of social interaction on the web. They are not plain reproduction or re-elaborations of paper-based genres, such as e-books, emails and online newspapers (Knox 2007, 2009a, 2009b). Even though there would be little agreement on the definition of such texts as *reproductions* of paper-based genres, we may well argue that the latter examples have a material counterpart in the physical world, in the form of books, letters and print newspapers. The texts analysed in this book, conversely, do not have a specific material counterpart in the physical world. A videochat cannot be equated with a phone call, as the former also includes the video *and* the opportunity to use text chat. The same may be said about blogging: a blog cannot be compared to a personal written diary, because among its constitutional features, we can list the distinctive information sharing and debating component via writing and other semiotic resources that all users have at their disposal. Social networking and media sharing communities are also distinctive web-based events that do not have specific counterparts beyond the web.

I have so far discussed forms of *stereotypical* spoken, written and digital discourse, all embedded in their traditional environment that can be provisionally postulated as, respectively, a spoken informal conversation, books and printed media, and loosely defined forms of web-based social practices. The materiality of *all* these texts is epistemologically different and it is precisely from the consideration of this *inescapable difference* that the book's rationale has been shaped. In my view, an overuse of the concepts of *hybridization* and *novelty* in Web 2.0 texts, heralded in many fields of studies (e.g., discourse analysis, multimodal studies, etc.), has been an escape route to bypass some theoretical and methodological problems, mainly concerned with the thorny issue of clearly defining and limiting the domain or object of study. Despite its undoubtedly efficient way of capturing the fluid complexity of contemporary texts, such a concept also runs the risk of muddying the water, thus making analysis practically impossible to replicate or contest. Conversely, construing models to interpret and *distinguish* among many linguistic and multisemiotic web-based events is a useful heuristic model to explore them, despite the fact that categorizations and taxonomies you will encounter in this book run the opposite risk of being excessively crude or naïve.

Since this project began in 2009, many things have changed from several standpoints. Technical affordances have increased the range of semiotic affordances accordingly (cf. the general improvement in web-based video communication ensuing in online, fluid communication that a smaller bandwidth made difficult or impossible), and genres have also started merging, as will be seen, for example, with blogs blurring into social content-sharing platforms.

The development of a unique method to explore such vibrant and protean texts and genres seems not only impossible, but also pointless. This is why I have used several methods to explore different aspects of the web-based world and also adapted them to fit the slippery digital environment as efficiently as possible. Multimodality has been the general umbrella rationale to draw together all these methods and theories of enquiry, as multimodality is, I believe, the most suitable theory *and* tool to draw them together consistently, while keeping them separate and distinct at the same time. Multimodality is a broad way of looking at communication, not only accounting for a set of communicative resources traditionally left outside mainstream linguistics, but also challenging and questioning what is traditionally *within* the field of enquiry of linguistics, as is the case of the speech and writing variation across digital media.

However, investigation into mode variation must also be grounded in linguistic theories, with the specific intent to preserve *language mode epistemological distinctiveness*. Biber established one of the soundest methods for analysing variation across speech and writing with his multifeature/multidimensional analysis (MF/MD, henceforth). However, some linguists, including Biber himself (Biber and Conrad 1999; Tribble 1999; Biber 2003; Scott and Tribble 2006) have argued that the study of lexicon may well produce similar effects to those produced by a more comprehensive investigation through MF/MD analysis.

In this book, a lexical approach has been adopted to investigate the verbal oral/written variation, analysing keywords and key clusters to scrutinize dimensions of variation. A key word is a word that occurs with unusual frequency, be it a positive keyword (occurring more frequently) or a negative keyword (occurring less frequently), in a text or in a collection of texts. It is, as such, a measure of statistical *unusuality*. A key cluster is made up of more than one word; it is a syntagmatic string of two or more word forms.

The notion of *keyness*, which has been central in recent studies in the field of corpus linguistics (Scott and Tribble 2006; Bondi and Scott 2010), has been typically intended as synonymous with text *aboutness*. However, this study contends that lexicon is also a strong predictor of language mode variation. Lexical studies can shed light on many kinds of differences that lie at the core of digital discourse as a whole.

Keywords can be defined, at their basic level, as *recurring words*. Recurring words have been mostly defined as indicators of *aboutness*, in other words of their prominence, for example exploring topics or stylistic aspects in literature. However, the fact that a word recurs or displays positive or negative keyness does not provide an analysis or an interpretation *per se*. Caution is needed in interpreting keywords, also because the reference corpus that is selected for a comparative analysis influences the results. This is why practitioners in the field suggest working with more than one reference corpus, as has been done in Chapter 3.

Special care is also needed when claims of statistical significance are made: they depend as well on how carefully the corpus is designed and built, for example taking into due account questions of representativeness. Despite these minefields, keyword and keyness analysis, alongside related topics, such as the exploration of lexical bundles (Biber and Conrad 1999), semantic prosody and semantic preference, are particularly interesting and relatively unexplored with regard to genre and register variation in digital domains. Many studies have investigated digital discourse and also variation across spoken and written language in different media, but lexical studies are still relatively unexplored out of the arena of *topics* and overall text *contents*.

To give a practical example, it is commonly held that writing on the internet is something very different from traditional writing. As I claimed in previous sections, writing is efflorescent in its rapid and editable nature over the internet. We all know that writing is different, it is different in how we produce and handle it in online processing: but is the lexicon a good indicator of this difference? This book suggests that it is. Keyness and lexical bundles should be accordingly partly similar to a spoken but also to a written reference corpus, according to this preliminary hypothesis. However, these indicators should not be considered as absolute or universal, but highly text dependent. They need be considered not only as genre dependent, but also as text dependent, as keyness is highly bound to text types.

The next section will discuss what key words and key clusters can tell us in the study of language variation, how lexicon is symptomatic in language variation and why it is useful to go beyond the single word (Wray 2002).

1.5.1 Word and Words

In studies on lexicon, the single word has always been a central feature of interest—for obvious reasons. The single word is perceived as the core unit of our lexicon and, as such, it has always enjoyed a privileged status also within the most common educational practices: learning words is, after all, one of the well-established practices of student learning strategies.

However, the word as a unit has come to be challenged by multiword units (or clusters or bundles), which are very common in English and also for learners of English: phrasal verbs, compounds, idioms and proverbs (McCarthy and Carter 2006). Other expressions are also very commonly used and taught in English, such as greetings and phatic expressions (e.g., *How are you doing? See you soon, you're welcome*), prepositional phrases (e.g., *in the street, at the hospital*) and common compounds (e.g., *travel agency, wheel clamp, driving licence*).

Studies into lexicon have promoted a better understanding of multiword phenomena. Following his early insights, Firth claims that the meaning of a word depends on how the word is placed within a text (i.e., *collocations*) and also on its inner lexical properties (Firth 1935), so that *dark* is part of the meaning of *night* and *vice-versa*, as is also evident in their high likeliness of co-occurrence in texts (Firth 1951/1957). Some pioneer linguists (Halliday 1961; Sinclair 1966) discussed the implications in the study of lexicon in the Sixties and collocation studies showed that a good portion of semantically transparent vocabulary is to a greater or lesser extent fossilized into restricted patterns (McCarthy and Carter 2006). As a consequence, the notion of collocation has been established as a central element and focus has been reallocated from one-word units to multiword units, considered as chunks of meaning, indispensable for communication, and especially so where interaction is concerned. The exploration into multiword phenomena has been made possible by the use of corpus linguistics applied to lexical studies that have revealed that much of our lexical production consists of multiword units and that human communication consistently relies on ready-made chunks of language.

Apart from the work by Sinclair, corpus-based work on grammar has yielded similar results. Biber, for example, calls these multiword units *lexical bundles* that are recurrent (i.e., establishing cut-off points) expressions, even though they may be semantically opaque and syntactically incomplete. This means that a bundle may be formed by a syntactically incomplete but meaningful expression, such as *be able to* or *a lot of*, but also of more functionally and pragmatically related expressions, such as *on the other hand*. According to McCarthy and Carter (2006), lexical bundles are "important structuring devices in texts and are register- (or genre-) sensitive" (2006: 10). Non-corpus-based research, such as phraseology, sociolinguistics and conversation analysis, has also emphasized the importance of multiword phenomena. Phraseology, for example, has shifted attention from fixed and opaque multiword units to other semantically oriented multiword phenomena. For example, as said, the notions of semantic prosody and semantic preference are paramount to shift attention from lexical to semantic areas of investigation (Sinclair 1996).

Likewise, from the perspective of text and textuality, words are not only the *substance*, the conceptual *subject matter*, but they also the purpose of shaping the text following organizational principles, for example of structuring information, as the notions of coherence and cohesion suggest (Halliday and Hasan 1976). In accordance with this, Bondi argues, "[T]he words and expressions that recurrently identify the conceptual structures and the organizational structures of a text or corpus can be studied to illuminate features of the discourse that produces the text or corpus. The keywords that point to the aboutness of a text or corpus will be key to the ontology of the discourse. *The keywords that point to textual organization will be key to the epistemology*" (Bondi and Scott 2010: 8, emphasis mine). The analytical software programmes used in this book, that is WordSmith 5 and 6 (Scott 2008, 2012), have been used to automatically retrieve keywords and lexical bundles, intended as recurring strings of multiword units in a wealth of web-based texts of various kinds, with the aim to explore language mode variation. In other words, a corpus-based lexical approach, among others, will reveal if and how speech and writing are integrated, assuming that different words and group of words, tend to co-occur differently in language modes. A multimodal approach will complement a purely linguistic approach to account for other significant semiotic resources.

The following chapter is devoted to the phenomenon of videochats. More than 300 hours of recording has allowed observations and hypotheses that will be detailed henceforward. Such data allow for qualitative analyses, and also explore users' demographic and qualitative interviews, with the aim of revising the use of semiotic resources in this specific interactional domain.

The linguistic phenomenon that has been explored with particular emphasis has been here defined *mode-switching* and represents the alternation between speech and writing in video conversations (Sindoni 2010, 2011a, 2012). Considering the novelty of the genre and amount of collected data, a fine-grained analysis has been carried out, studying the main interactional patterns, both non-verbal and verbal, with particular reference to mode-switching. This is why a corpus-based lexical approach has been applied more systematically in Chapters 3 and 4, respectively, devoted to blogging and media sharing communities, since some aspects, such as context, localization, provenance and authorship were less evident, and in some cases, impossible to monitor.

In conclusion, speech and writing appear as complex language modes, revealing to a significant extent the nature of human relationships, be they personal, social, political, cultural in very broad terms and in different ages and cultures. Speech and writing have always shaped the history of communication, comprehending domains such as education, power, solidarity, intimacy, subjugation and cultural inclusion and/or displacement.

A multimodal approach, informed by other linguistic methodologies, mainly including corpus-based lexical studies, conversation and discourse analysis, has been adopted in this book to embrace these phenomena as comprehensively as possible, in the attempt to capture the evanescence of the digital spoken/written divide, at least for the time I am writing and you are reading this book.

2 Spontaneous Web-based Video Interactions

2.1 REINTERPRETING SEMIOTIC RESOURCES IN THE LIGHT OF THE VIDEOCHAT

What is a videochat? It is commonly defined as a web-based system connecting two or more people located in different places to communicate through online, simultaneous two-way audio and video transmission. One user interviewed for this book, simply said that "it is a way to videocall friends and people and feel *like you really are* with them" (emphasis mine). This chapter addresses the question of defining videochats by developing a descriptive model within the book's broader agenda of tackling multimodal, multisemiotic and multicultural web-based interactions.

The first section of the chapter provides a theoretical introduction to the study of videochats, seen as dynamic texts that create a new form of communication. Social semiotic resources are arranged in such innovative ways in videochats that they produce a new terrain for investigation for researchers in the field of linguistics, communication, media studies, visual ethnography and digital literacy. The chapter singles out aspects relevant to the study of videochats, such as the alternation of speech and writing, new patterns in proxemics and kinesics and impossibility of eye contact, which are so striking as to warrant special investigation. In particular, speech and writing are technologically integrated, allowing participants to *mode-switch*, that is, to alternate between spoken and written discourse (Sindoni 2010, 2011a, 2012). New arrangements of verbal and non-verbal resources attempt to simulate face-to-face conversations. However, the illusion of a face-to-face conversation dissolves as soon as videochat-specific resources are unpacked. Despite the growing research into non-verbal behaviour, videochat data challenge visual analysts and researchers for a variety of reasons illustrated in this chapter.

These theoretical observations require empirical support provided in the second part of the chapter. Analyses and observations are based on two separate datasets: a survey that reports data from interviews on videochat users' self-perceptions (Sindoni 2011a), and a corpus consisting of more than 300 hours of recorded spontaneous web-based interactions from multi-room videochat clients, for example, Camfrog. The survey provides a preliminary description of users' self-perceptions as a background through which to subsequently engage with video data. These datasets point to re-current patterns and interpretative models are suggested.

The survey provides some insights into relevant issues, such as interview-ees' use of video/text chat and preferred mode of interaction. The data are consequently organized to account for the interviewees' self-perceptions on mode-switching during a videochat and other issues, following criteria discussed in the first part of this chapter. The corpus gathers video data, and in the subsequent subsections the rationale for the development of a transcriptional model is discussed, building on the theoretical bases pro-vided. The multimodal matrix presented in Appendix 2.1 is the result of the transcription of a multiparty videochat, selected as a case study. The notion of mode-switching is foregrounded, in particular in relation to the switch-ing between spoken and written discourse. This event is viewed within the context of more complex multisemiotic activities and interactions.

Reflections and conclusions are drawn vis-à-vis the contribution that intersemiotic studies can potentially provide for web text analysis, given the constant expansion of web practices and the challenge it brings as regards new notions of textuality. In this sense, videochats raise issues and philo-sophical questions addressed in the concluding remarks.

The following sections suggest ways in which videochats conform to, but also disrupt, our perception of many of the traditional categories on which multimodal analysis is based. New technologies pave the way for new forms of interaction and with it a world of paradoxes, illusions and deceptions. Below, a specific text genre is explored in this deceptive, illu-sive and paradoxical light: even as it mimics face-to-face conversations, the videochat distances itself from *in praesentia* interactions.

2.1.1   Speech and Writing in Videochats

In videochats, participants may use speech and writing at the same time, conversing but also writing and sending comments. They *may* and, in fact, often *do mode-switch*, that is, alternate between speech and writing and, in so doing, they are able to mix both modes in personal ways and achieve specific communicative purposes by drawing on both resources at the same time (Sindoni 2010, 2011a, 2012). Mode-switching is a descriptive and exploratory label that takes this communicative possibility into account. It paraphrases the linguistic notion of code-switching, which is generally held to mean the use of more than one language or variety in a conver-sation (Heller 1988). It also draws on Halliday's concept of written and oral modes of communication, whereas mode is exclusively taken to refer to *verbal language* (Halliday 1978; Halliday and Matthiessen 2004). It is based on the observation that the web has so far developed a wealth of new genres that intertwine speech and writing in unparalleled ways. Mode is here referred to exclusively as *the alternation of language modes, that is, speech and writing*, thus excluding the alternation within the whole array of other semiotic resources that will be dealt with in the subsequent chapters.

Videochats basically rely on speech technologies, but there are constant alternations between written chat and audio/video conversations, especially in more informal contexts. In other words, an online conversation may be interspersed with written comments.

Preliminary research questions include: what drives users to associate oral and written technologies of the word—to borrow well-known categories from Ong (1977, 1982)—in videochats? Is it because they are just there or is there something else behind, or perhaps beyond, the mere exploitation of the full technological affordances available in videochats?

More than two decades ago, Halliday argued that the categories of "written" and "spoken" are indeterminate:

> [T]hey may refer to the medium in which a text was originally produced, or the medium for which it was intended, or in which it is performed in a particular instance; or not to the medium at all, but to other properties of a text which are seen as characteristic of the medium. ([1987] 2002: 335)

Hence, the need to specify which discourse variable is being referred to when a given mode is selected. As a working hypothesis, in the case of videochats, the videotrack may be considered as the main online conversation, while written comments might be defined as integrations of the main text. However, many instances of videochats can be defined as the exact *opposite*: written comments constitute the main text, interspersed with occasional video/audio comments. The great variability in these integrations needs to be further explored within a purposely created, but empirically validated, theoretical model.

As Table 2.1 shows, Halliday identifies distinct arrays of features for oral and written modes, highlighting the fact that both modes are equally systematic, organized, regular and productive of coherent discourse (Halliday [1987] 2002: 340). Given the prominence of grammar for writing (theories on grammar were written in books, and, as such, they were typically based on written discourse), Halliday was at pains to create theoretical bases for a grammar that specifically took spoken resources in the linguistic system into account.

While speech exhibits higher levels of grammatical intricacy, writing displays a higher lexical density. So Halliday claims that both modes are equally complex. Instead of trying to indicate the primacy of one mode over the other, the question is how to define and identify kinds of complexity: occurrences and degrees of overlapping between the two modes (Tannen 1982a). In the examination of clause complexes, for example, concepts such as embedding and hypotaxis must be analysed separately. Halliday considers this strategy as fundamental "to do justice to the particular mode of

Table 2.1   Spoken and written discourse features

Speech	Writing
Spun out	Dense
Flowing	Structured
Choreographic	Crystalline
Oriented towards events (doing, happening, sensing, saying, being)	Oriented towards things (entities, objectified processes)
Process-like	Product-like
Intricate	Tight
With meanings related serially	With meanings related as components

Adapted from Halliday ([1987] 2002).

organization of both spoken and written discourse" (Halliday [1987] 2002: 343). He also points out that "grammar needs to distinguish between the constituency relation of embedding, or rankshift, where one element is a structural part of another and the dependency relation of 'taxis,' where one element is bound or linked to another but is not part of it" (Halliday [1987] 2002: 343–44).

In other words, only a grammar that considers cultural, social and linguistic distinctions between speech and writing fully accounts for differences and overlaps pertaining to both modes (Halliday [1987] 2002), though we would add that the multisemiotic and multicultural framework proposed here is the logical progeny of Halliday's thinking. A comparative analysis of the two different types of texts, oral and written (audio/video and written text), serves the purpose of monitoring and includes *all* possible convergences, discrepancies and decoupling, given that writing during a videochat will show features characteristic of speech. However, a more quantitative method (i.e., corpus linguistics) will be applied in the following chapter, whereas in this chapter qualitative observations will be reported and discussed, considering that amount and kind of video datasets require a more comprehensive approach. Furthermore, the question of mode-switching will be placed in the foreground, leaving other, corpus-based lexical observations for the subsequent chapters. Video datasets dealt with in this chapter will be treated with methods from conversation analysis that is a suitable theory and tool for analysing web-based, live and spontaneous interactions.

Obviously, recordings, which provide access to the synchronization of the videotrack with written comments, are an indispensable tool to this end. One working hypothesis is viewing written comments as an innovative use of speech (cf. Bader 2002; Schachtebeck 2005), which may be provisionally called *graphic speech*, as features of this mode may be associated with spontaneous oral interaction.

Videochats simulate face-to-face conversations, giving the illusory perception of sharing the same context of situation, a visual and psychological perception created virtually through a range of different strategies. A fundamental factor adding to the illusory perception of face-to-face conversation is that verbal exchanges occur in real time. Moreover, users also see each other in real time, so that conversations reproduce long or short verbal exchanges *in praesentia*, with turn-taking, pauses, hesitations, and so on, respecting social roles and cultural conventions. However, this system of interaction is a rather crude imitation of what happens in face-to-face conversation. First and foremost, participants do not share the same context of situation. Just as in face-to-face conversation, they take turns, although now they are physically separate and what they see is a projected image of their discourse partner. Videochat software systems commonly provide a larger-scale close-up of the discourse partner located in the foreground, while one's own smaller-scale close-up image appears at the bottom or at the top of the frame.

Moreover, the size of one's own image and that of one's discourse partner varies. In *Skype*, the former is smaller and usually placed below the videochat window, while the latter is larger and set in the foreground. As another example, the *MSN* videochat interface features a similar arrangement of frames. The rationale for this hierarchy of images is self-evident: interest is mainly focused on the discourse partner, but one's own image is nonetheless there, making oneself aware of one's own body movements, posture, and so on. In the *MSN* chat interface, both participants have an icon associated with their account and both of these appear in the chat window.

From an intersemiotic perspective, videochats thus include two differently integrated frames showing live images of users. Contrary to what happens in face-to-face interactions, while having a video conversation, discourse partners have the unique opportunity to see themselves. Observing oneself during a conversation produces a series of psychological effects influencing the verbal and non-verbal behaviour of online exchange. In other words, casting a sidelong glance at oneself during a conversation may change, if not determine, the way one speaks, gesticulates, smiles and so on; it entails adjusting to the addressee's expectations, but mostly adjusting to one's own ever-changing expectations, as well (Sindoni 2012).

2.1.2 Social Distance in Videochats and Their Place in the Study of Proxemics

Perception, use, structuring and management of space all have the effect of framing social expectations, interactions and ways in which actual communicative events take place. Social interactions, at least when defined by face-to-face communication, occur in a setting, which is physically perceived by participants, who negotiate their relationship with space and with one another in meaningful patterns. Hall (1966) called this area of interest "proxemics," distinguishing four different social distances (i.e., intimate, personal, social/consultative and public).

Social distance in videochats is fixed, and participants stand in a frozen space. This means that medium constraints (close to very close shots of the projected image of both participants) do not allow mutual negotiation of space. In other words, distance is not established by those who interact, but between one participant and one machine. This distance foregrounds the *representation* of distance among users. Since space reflects environmental arrangements that may encourage communication (i.e., sociopetal space, Hall 1966) or keep people apart (sociofugal space, Hall 1966), data need to be gathered in order to decide whether a videochat fits into the first or second category. Furthermore, studies on proxemics have come to include postural identification (i.e., sitting, standing), distance, frontal orientation, and body positioning (Harrigan 2005: 149); none are ever easy to discern in videochats, especially where the bodies of participants are only partially visible or fall outside of the projected frame.

Ways of managing space during multimodal interaction affect the ways in which communication is organized overall and carried forward (Heath and Luff 2000). Notions of territoriality may well refer to the act of laying claim to, and defending, a territory from political, ideological or cultural invasions, but may also refer to personal spaces, whose "invasion" may be perceived as equally, if not more, threatening. Hall and Hall (1990) argue that space is perceived by all senses, not by vision alone: "auditory space is perceived by the ears, thermal space by the skin, kinesthetic space by the muscles, and olfactory space by the nose" (Hall and Hall 1990: 11).

However, proxemics may also be interpreted in broader terms, for example, taking into account how historically people have come to negotiate space with other people in public places and how notions of personal space have shaped, related or opposed notions such as those of neighbourhood, solidarity, estrangement and alienation. Simmel was one of the first to comment on how social intercourse in modern times is more based on the sense of *seeing* than on the sense of *hearing* ([1908] 1921). At the beginning of the twentieth century, he was able to articulate the emotional breakdown occasioned by the irreversible loss of connection experienced by modern people in the following terms: "[and this] brings us to the problems of the emotions of modern life: the lack of orientation in the collective life, the sense of utter lonesomeness, and the feeling that the individual is surrounded on all sides by closed doors" (Simmel [1908] 1921, quoted in Emmison and Smith 2000: 192). Sennett contrasts past and present times. In the former, people used to seek rewarding contact with strangers; in the latter, strangers are perceived as threatening. "In the mid 19th century [. . .] there grew up the notion that strangers had no right to speak to each other, that each man possessed as a public right an invisible shield, *a right to be left alone*. Public behaviour was a matter of observation, of passive participation, of a certain kind of voyeurism. The 'gastronomy of the eye,' as Balzac called it" (Sennett 1977: 27, emphasis mine). The *right to be left alone* is taken for granted in contemporary times. However, sociability takes on different articulations in urban and rural areas. Cities are considered hostile and unwelcoming places, where anonymity is the rule. According to Lofland (1973), coping with the city means coping with strangers.

Goffman defines personal space as the "space surrounding an individual, anywhere within which an *entering* other causes the individual to feel encroached upon, leading him to show displeasure and sometimes to withdraw" (1971: 29–30). According to Emmison and Smith (2000: 211), the works of Simmel ([1908] 1921, 1949), Lofland (1973) and Goffman (1971) highlight "the importance of display in everyday social life. They suggest that people work to give off meanings and impressions. These facilitate everyday social interaction by providing visual clues to others as to who one is and what one is doing. Aside from gaze, the body provides one of the major tools that people have for accomplishing this complex task." Social interaction is thus intertwined with proxemics, the latter broadly interpreted as a meaningful way to establish and maintain relationships. The body helps to achieve this communicative task, for example by directly modifying (e.g., tattoos), by adorning (e.g., clothes) and by moving (e.g., movements, posture, etc.; cf. Emmison and Smith 2000: 212). The human body itself is a semiotic text, as it may be described as an orienting system of signs and communication (Baldry 2011, personal communication).

In videochats, the frame delimits one's personal space, and that of others, and the distance between oneself and the other participant is fixed and is completely deprived of social connotations. No matter what social variables are at stake, video-based interactions occur within a fixed frame, which seems to erode culture-bound proxemic patterns. In other words, regardless of social and cultural positions, videochat communication systems erase all kinds of differences and favour a flat representation of social and cultural identities.

Management of space in videochat is problematic, because it *stages* a spatial setting and a spatial distance. Participants can, in fact, provide a more or less distant image of themselves, for example displaying a very close shot or a medium shot to their discourse partner/s. This may be achieved through positioning oneself vis-à-vis the webcam, which is never neutral, but is an integral part of the interaction, in that it may represent an *intentional act* on the part of each participant.

When users decide to project their screen, activating the video function, they decide *what* to project. Each choice is significant from a semiotic standpoint. In many instances, the choice to use the video function does not equate with projecting one's own image. Users can send pictures, animations, written messages or slideshows. The video image projected from the machine is a potential meaning-making event that can be used to good effect. This is related to proxemics, because it depends on how social space is constructed and negotiated within the environment. All that is embedded in the screen contributes to shaping a represented social distance. Frames can include a frontal or oblique angle, the latter usually showing a partial view of participants. Perceived proxemics can also be altered by webcam position, that can be placed at a distance so as to represent "long shots," thus potentially including other subjects, who may happen to be within the webcam's reach and who are often unaware they are being shot.

The frame can also be "empty," that is, not including *any* participant, but reflecting parts of the environment, such as walls, desks, the bed, and so on. When prompted to explain the reason for such apparently arbitrary frames, my informants justified their choice by defining these video frames as a "guarantee of reality," that is, to show other users that "there is someone out there, even if you don't see them right now," as one of the informants contended.

2.1.3   Framing the Body in Videochats: Reinterpreting Kinesics

Just as proxemic patterns are linked to culture and individuals and any visual analysis needs to take such variables into account, so the same may be said about kinesics. Very few body movements have invariant meaning within or across cultures, because body movements cannot be translated as directly as can verbal behaviour (Kendon 1990, 2004). Apart from messages conveyed by movements, the additional question of intentionality must be taken into account (Norris 2004). In verbal communication, there is a *deliberate* attempt to convey a message to a recipient. With non-verbal behaviour, the question of intentionality is less clear-cut, because some actions may well be defined as intentional (e.g., deictic gestures, such as pointing to objects and people in the physical world, Norris 2004: 28). Others are halfway between intentional and unintentional (e.g., iconic or metaphoric gestures, possessing a pictorial content and mimicking what is conveyed or hinted at verbally, Norris 2004: 28). Others are unintentional (e.g., postural behaviour, Norris 2004: 24–27). However, these generalizations do not purport to represent stable relationships between intentionality and behaviour. Any specific interaction may present different configurations of intentionality from those sketched here, as any interaction is culturally, socially and individually determined. However, the question of intentionality is seminal in any visual exploration of body movements.

Kinesics has traditionally focused attention on hands and head movements (Kendon 2004; Martinec 2004), which have been studied as "action" behaviour, which consists of discrete units of analysis, including "onset" and "offset" points (i.e., distinct beginning and ending of actions). Some studies focus on verbal and non-verbal congruence, analysing speech units, pauses and timing of body movements (Boomer and Dittman 1964), whereas social psychologists are more concerned with inner states revealed by non-verbal behaviour (cf. Ekman and Friesen 1969). However, multimodal interaction does not deal with what people are thinking, but with what people are *communicating* in interactions (Norris 2004).

In the context of videochats, body movements cannot be completely ascertained by analysts, since hand movements, for example, may be semivisible and, as mentioned above, partly fall outside the video frame. Other body movements, such as postural shifts, may even *totally* fall outside the video frame and, as such, can be missed both by the other discourse partner and subsequent observers (e.g., multimodal analysts, researchers). Video frames do not include the entire body, but usually frame the face, and optionally a part of the torso, creating a *frozen yet living* image of users. The *frozen yet living* definition attempts to capture the contradiction in the video frame which includes a frozen, fixed and partial representation of the body and at the same time presents a living representation of speech, facial expressions and body movements. The latter are typically instantiated by hands and head movements and torso postural shifts. The discourse partner's face is, thus, a metonym for the body realized by another face: one's own. This body fragmentation and dematerialization will be discussed later.

Participants can employ gestures for many other communicative purposes. For example, when engaged in playing quizzes, they often write numbers on their hands and display them on screen to give the exact answer to a quiz, as it may be difficult to have the floor for speaking. Considering the familiarity that users can achieve by participation in such communities, they are also very likely to perform a wide range of everyday activities that imply a certain level of acquaintance, and, in some cases, of familiarity with other members, for example yawning, drinking, smoking, eating, phoning, stretching, scratching, texting, combing hair, even sleeping.

2.1.4 The Illusion of Eye Contact in Videochat and Its Position in Theories of Gaze

Gaze is one of the most effective resources in interpreting and making sense of a discourse partner's attitude, stance and behaviour. Early research on visual behaviour was concerned with a series of questions on different levels of communication:

> 1) What does the behavior tell a trained (or untrained) outside observer about the subject? 2) How is the flow of conversation regulated? 3) What is the looker looking for? 4) How does gaze direction influence the receiver? 5) What attributions does a receiver make on the basis of the other person's visual behavior? 6) To what extent is visual behavior "communicative" in the narrowest sense set forth above? (Ellsworth and Ludwig 1972, in Mortensen 2009: 248)

As a dependent variable, gaze has been used to measure stable individual and group differences, the regulation of conversation flow and the search for feedback in interaction. As an independent variable, it has been shown to influence emotional responses and cognitive attributions (Ellsworth and Ludwig 1972). In the context of videochats, significant questions may be identified in items 4 and 5 in the list quoted previously: how gaze direction influences the receiver and the meanings attributed to the other's visual behaviour are particularly significant to this discussion.

Studies in psychology, interactional analysis and ethnography of communication have focused on gaze and its role in interactions, and have provided insights into the different levels of analysis and stratification of gaze functions. The seminal paper by Argyle et al. (1973) illustrates the different functions of gaze: (1) information seeking (i.e., obtaining immediate feedback); (2) signalling either interpersonal attitude or prosodic accompaniment of speech; (3) controlling the synchronization of speech; (4) the relationship between mutual gaze and intimacy; and (5) the inhibition of gaze (i.e., avoiding undue intimacy). These separate functions were identified and cross-checked in their experiment with participants acting in different looking conditions (i.e., seeing vs. not seeing the partner), confirming hypotheses about the amount and distribution of looking for each function. This model of categorization in terms of functions allows for a systematic analysis of gaze in keeping with different interactional conditions. In other words, interactions are context-dependent and although some aspects are more likely to occur in interaction (e.g., obtain feedback), others may be absent or mutually exclusive (e.g., the relationship between gaze and intimacy vs. inhibition of gaze to avoid undue intimacy). Any analysis concerned with gaze is, thus, heavily dependent on the context in which communication occurs and while some aspects may be ignored, others need to be carefully taken into account (Harrigan 2005).

To what extent is turn-taking accompanied by gaze in videochats? Sacks, Schegloff and Jefferson (1974) were the first to find no ideal overlap in conversation and identified the *transition relevance place* (i.e., the end of a turn construction unit in conversation analysis) as the crucial point at which floor changes are negotiated.

Goodwin (1980, 1981) and Kendon (1967, 1990) demonstrate that turn-taking behaviour in conversation is mainly regulated through gaze. Kendon shows that transition relevance places are anticipated through gaze, whereas Goodwin argues that having listener gaze is so important for the turn-claim process that turn claimants will restart their utterances until the speaker's gaze assures them that they *do* have the floor. Telephone and computer-mediated communication present problems in turn-taking that resemble those of the differences described in blind and sighted interaction, since gaze is a fundamental facilitator (Everts 2004). Research shows that subjects use more turns when they experience more gaze (Kendon 1967), and experiments have been carried out to show effects of eye gaze on mediated group conversations (Vertegaal and Ding 2002).

Unlike telephone conversations, which do not provide eye contact cues, videoconferencing systems give the incorrect impression that the remote discourse partner is avoiding eye contact (Yang and Zhang 2001). Some videoconference systems have onscreen cameras that reduce the difference in parallax (Otto et al. 1993). This issue is also being addressed through research that generates a synthetic image with eye contact, using stereo reconstruction or cutting-edge systems able to simulate eye contact between a three-dimensionally transmitted remote participant and a group of observers in a 3D teleconferencing system (A. Jones et al. 2009).

Management of gaze is bound by many factors, such as context, culture, media and so on and, as such, predictions on its use cannot claim any kind of universality. However, some uses of gaze may be intuitively and successfully interpreted, especially in conversation. Eye contact is in effect an almost universal feature of face-to-face interaction and, even though direct gaze in some cultures is permitted only in intimate contexts, exchanging looks, staring and eye contact are natural properties of spontaneous conversation.

Figure 2.1   Impossibility of eye contact

This feature is problematic in videochats and a fundamental question to be asked is how analysts can analyse gaze in spontaneous but technologically mediated conversation.

Eye contact is impossible in videochats, because participants *either* look at the frame where the discourse partner's image is projected *or* at the webcam, thus giving the other the illusion of direct gaze. However, in videochats, virtual gaze is *never* reciprocal: if Participant A looks at the webcam, Participant B feels she or he is being looked at straight in the eyes. But this perception cannot be reciprocal, since Participant B cannot reciprocate gaze *at the same time* (see Figure 2.1).

Giving the other discourse partner the impression of direct gaze paradoxically excludes the possibility of seeing the other *tout court*. The problem is solved partially, only when the webcam is in the centre of the screen (see Figure 2.2). However, not even embedded webcams allow perfect eye contact. Simulating eye contact in virtual worlds is far from easy.

Vector analysis is thus complex, if not impossible, because the impossibility of eye contact alters traditional parameters of analysis. Gaze is thus still a facilitator for conversation in videochats, but its incidence and role is not easy to gauge. Should gaze be absent, as in the case of telephone

Figure 2.2   Eye contact simulation via embedded webcam

conversations, other facilitators, such as prosodic elements, take over, but the peculiar presence of indirect and nonreciprocal gaze requires rethinking to be undertaken of well-established models of visual analysis (van Leeuwen and Jewitt 2001; Sindoni 2012).

The well-known distinction between management of gaze on TV and in the cinema may be recalled to highlight how these media genres are epistemologically different. As in videochats and in television newsreading, looking at the camera simulates direct contact with the audience, while actors never look at the camera, to avoid breaking a fundamental rule, which, by definition, excludes virtual eye contact between characters and audience (Metz [1980] 1993). The depicted world is a separate universe, living an inclusive, realistic life; the audience can identify and enter this world precisely because of its separateness. If characters looked directly at the camera, thus erasing this ontological separateness, they would spoil the audience's voyeuristic pleasure, as they peep into this fictional world without being seen. This is not always the case in television, where the initiators of communication, for example TV newsreaders, need to involve the audience, simulating contextual presence. As Kress and van Leeuwen point out, there may be a "demand" or "offer" in both static and dynamic images: "the participant's gaze (and the gesture, if present) demands something from the viewer, demands that the viewer enter into some kind of imaginary relation with him or her. [. . .] [The offer image] 'offers' the represented participants to the viewer as items of information, objects of contemplation, impersonally, as though they were specimens in a display case" (Kress and van Leeuwen 2006: 118–19). In addition to this, they further point out that at stake is the question of the communicative power or "entitlement," as Sacks (1992) calls it, as regards everyday face-to-face communication (quoted in Kress and van Leeuwen 2006: 121). The current impossibility of eye contact in videochats implicitly underscores the virtual conversational nature of videochat.

2.2 A SOCIOLINGUISTIC SURVEY INTO SELF-PERCEPTIONS IN VIDEOCHAT USE

Qualitative data have been gathered through a survey sampling one hundred English-speaking high school and university students that includes questionnaires, and structured and semi-structured interviews conducted over a period of nineteen months (February 2009–September 2010, cf. Sindoni 2011a). The data suggest that various heuristic categories may be devised to account for users' self-perceptions with regard to mode-switching. Even though users were not able to quantify mode-switching, not even in rough terms, what has emerged is a significant recourse to the writing mode in casual videochat conversations. Further details are given in the following sections, where users' interviews are discussed in greater detail.

2.2.1   Method and Data

The following analysis is based on a sociolinguistic investigation into teenagers' engagements with videochats. The data include questionnaires, and structured and semi-structured interviews conducted over the course of nineteen months. The one hundred interviewees ranged in age from fifteen to twenty-five, although most were in the mid-teens to early-twenties range. Most of the interviewees were from the United States, twenty-two were from Great Britain, ten from Australia and seven from Canada. Formal interviews lasted from one to four hours and included a pre-interview survey on the interviewees' demographic background and media use. The preinterview step was conducted on a larger sample of informants, contacted via a purposely created Facebook discussion group of videochat users. The sample of one hundred informants was created after collecting data on more than 600 subscribers to the discussion group, on the basis of their web literacy, age, first language (i.e., English) and willingness to join the project. Table 2.2 reports the distribution of interview administration methods.

Of the interviewees, 62 percent were female, 38 percent male in the fifteen to twenty-five year-old age range, all attending high school or university. After the pre-interview survey stage, they were asked some questions as regards their frequency of videochat use (Figure 2.3), use of speech and writing (Figure 2.4), and further semi-structured questions about why they

Table 2.2   Administration methods of structured and semi-structured interviews

Videochat	42
Instant messaging chat	25
Face-to-face	20
Phone	13

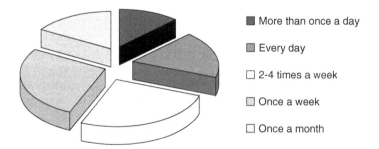

How often do you use a videochat?

- More than once a day
- Every day
- 2-4 times a week
- Once a week
- Once a month

Figure 2.3   Frequency of use

resorted to writing while videochatting (Figure 2.5). Another separate set of questions was devised to explore the practice of looking at oneself during a videochat, also scrutinizing interviewees' feelings and expectations on this matter. Figure 2.3 shows that the selected interviewees were all active videochat users. In their interviews, a significant majority showed mastery of the medium and full awareness of the medium's affordances. Thirty-six percent of interviewees claimed they used more than one videochat software programme, and commented on their familiarity with a wide range of video-based interactions. "Once" and "two-four times a week" options amount to 52 percent.

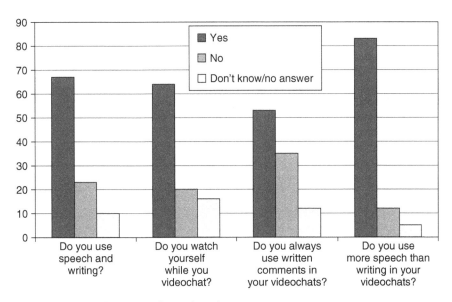

Figure 2.4   Combination of speech and writing

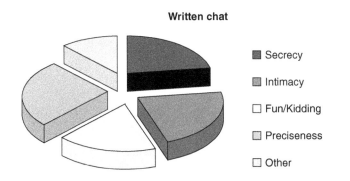

Figure 2.5   Reasons why users employ written comments during a videochat

Table 2.3   Summary of explanations on each macrocategory provided by interviewees

1.	Secrecy	Do not want to be heard/ tell a secret
2.	Intimacy	Communicate particularly intimate feelings (e.g., love, friendship, loyalty, etc.) or express face-threatening feelings (embarrassment, shame, taboos, etc.)
3.	Fun/Kidding	Add visual (e.g., emoticons) or expressive (e.g., swear words) power to communication
4.	Precision	Give precise information (e.g., addresses, telephone numbers, names, references, etc.)
5.	Other	No reason given as to why they resort to writing

Figure 2.4 reports data collected during structured and semi-structured interviews. The first part of the interview included questions reported in Figure 2.4, dealing with the use of speech and writing, the practice of watching oneself during video-based interaction, frequency of the use of writing, and mode preference/use.

The data indicate that videochat users tend to employ both speech and written comments, but that speech is, not surprisingly, the preferred mode of interaction (83%). Writing, nonetheless, was part of these communicative exchanges, with the constant use of written comments, as 53 percent of interviewees claimed that they used text chat "on a regular basis." The use of writing was recognized by informants as more "contrived," "useless," "artificial," "boring" than the use of speech, but even so employed for a number of reasons. Interviewees showed a lack of awareness when asked about their specific use of writing in videochats. However, when prompted to think about the purpose of written comments in specific examples, they gave a number of answers. Details are given in Figure 2.5.

Figure 2.5 displays four main categories that describe interviewees' answers, plus an extra category where answers that did not fit into the previous four categories are included. However, these categories are rather crude abstractions that cannot account for the processes, contexts and situations described during the interviews. Accordingly, Table 2.3 elaborates on the data given in Figure 2.5.

2.2.2 Discussion of Results

Table 2.3 provides revealing insights into the overall characteristics of videochat interactions in this age group. Thus, in Category "Secrecy," users reported situations where they used text chat to avoid being heard by third parties (e.g., parents/siblings in their room) not involved in the conversation. Written chat was identified as a useful strategy to communicate directly with the addressee without being heard by any other person, although this category makes sense only in the context of one-to-one videochats. In Category "Intimacy," users discussed particularly intimate contexts where they felt "emotional" for some reason, so written chat proved a "safer" way to express confidential or private feelings (62% reported they used text chat during a video conversation with a boyfriend/girlfriend). Category "Fun/ Kidding" is focused on visual items that users select at certain points of the interaction to enrich the semiotic resources during the communicative event. Category "Precision" accounts for situations where users did not want to be misunderstood and resorted to writing in order to give precise information, such as numbers, figures or proper names. Other categories could be devised to account for other situations, for example when technical problems prevented speakers from hearing each other. Writing was the obvious solution when tackling such technical issues.

In educational environments, Categories "Intimacy," "Fun/Kidding" and "Precision" are especially interesting for researchers, teachers and practitioners in the field of e-learning, distance learning and digital literacy. Category "Intimacy" suggests possible strategies to avoid or reduce face-threatening feelings, which hinder motivation and may block successful learning, particularly in some cultural contexts. Category "Fun/Kidding" opens up the way for a multimodal approach to learning, where the visual dimension is given equal importance as the verbal (i.e., written or oral) dimension (Kress and van Leeuwen 1996, 2006). The opportunity to use visual messages (e.g., emoticons, pictures, drawings) along with spoken and written messages may encourage the inclusion of visual resources in learning activities. Category "Precision" has more than one implication. For example, it may be useful in a teacher-led video-based interaction, where the teacher may send any kind of written information to clarify or specify difficult items (e.g., spelling) for the learner. Written comments may also be used to check understanding or to ask for feedback.

This survey does not claim statistical representativeness, but is meant to provide some qualitative data on users' self-perceptions. As the word itself suggests, "perceptions" do not always equate with real use and behaviour, but these perceptions nonetheless give insights into the possible *patterned* and *ritual* use of digital/video interaction, which can then be described in terms of that which Goffman called the differences between "system properties" and "ritual procedures" (1981). Without going into detail on Goffman's work, we may observe that videochats are examples of "focused interaction" and display all those features typical of focused interaction. System properties have to do with elements ensuring intelligibility, such as orderly turn-taking, while ritual procedures are concerned with different properties, such as face protection, politeness, and so on. Indeed, interviewees showed a consistent penchant for patterned turn-taking that followed slightly different patterns and prompted mode-switching. The sample interviewed does not allow for generalization, but face protection was indicated as being performed preferentially by the written mode (78%).

Face "is something that is emotionally invested, and that can be lost, maintained, or enhanced, and must be constantly attended to in interaction. In general, people cooperate (and assume each other's cooperation) in maintaining face in interaction, such cooperation being based on the mutual vulnerability of face" (Brown and Levinson 1987: 66). For example, a female interviewee claimed: "I use text chat when I feel my friend is being embarrassed or when I feel awkward myself, putting those funny lines on the screen or . . . don't know, those smileys and the like." In another interview, a male interviewee argued, "when it comes to personal matters, I write a line or two to my girlfriend."

In the next sections, field observations on multiple-room videochats provide raw data; subsequent analyses will add another standpoint to the issues being discussed so far.

2.3   A MULTIMODAL ANALYSIS OF MULTIPARTY VIDEOCHAT INTERACTIONS

What follows is a study based on about 300 hours of recorded spontaneous interaction, adopting a multimodal approach to a multiple-room videochat, namely Camfrog (http://www.camfrog.com), and using conversation analysis methods to address specific interactional patterns within the community.

Data were recorded at different times during the day and night, in order to include a varied range of interactions, allowing comparisons of what occurred at different times. Camfrog is a multilingual videochat client functioning as a standard instant message programme, in that it allows users to instant message each other on a private basis. Participants can also interact via a private one-to-one audio/videochat. One difference with most other instant message programmes is that users can also connect to one of thousands of chatrooms available, view other users' live video, use text and audio chat. Its main functionalities include: (1) *chat*: users can access Camfrog hosted and user-hosted videochat rooms, create their own videochat rooms selecting themes and setting up customized topics and rules, provided that they follow the basic Terms of Service; (2) *file sharing*: users can send files to each other, thus increasing the integration of multimodal resources working at the same time to produce, interpret and exchange meanings; (3) *snapshots*: users can take pictures through their webcam and send them. This functionality adds a static visual resource to the dynamic visual resources deployed during a videochat (i.e., live interaction); (4) *profiles*: users from age sixteen and older can create personal profile pages; (5) *mobile version*: Camfrog can be used on mobile devices for one-to-one private audio calls; and (6) *watch or broadcast videos*: users can watch or broadcast fast and high-quality video streams.

Camfrog client integrates multiple windows. A top-left banner features one's own projected image and other participants' projected images on the right. The number of users that it is possible to see at the same time depends on possession of the free or commercial (i.e., Pro) Camfrog version. Users may click on the list on the vertical right banner to see other participants, whether they are talking or not. Not all users are equipped with a webcam, so a micro-webcam icon on each user's name allows others to see whether the user may be seen or not. Camfrog also features written chat and users may draw on different fonts, sizes and colours to customize their written exchanges. This practice is common to any video or chat software client, and is a well-established visual system in web-based interactions. Users have access to both spoken and written resources. Clicking the "Talk" icon enables users to speak. Their nickname appears on the right of the "Talk" icon to avoid confusion, in order to facilitate identification with the person holding the floor. Besides providing a list of online participants, the interface offers a view to online users. Users are divided into three categories: visible, hidden and lurkers. Lurkers are online but are not interacting; they "spy" on active users, like silent co-participants.

Resource integration and meaning-compression principles are at work. The resource integration principle "refers to the ways in which the selections from the different *semiotic resource systems* in multimodal texts relate to, and affect each other, in many complex ways across many different levels of organization" (Baldry and Thibault 2006: 18). Ensembles of semiotic resources thus produce effects that differ from those produced by a single semiotic resource *and* from the mere *sum* of semiotic resources. Unpredictable effects are therefore originated when a set of different semiotic resources are conflated in a single meaning-making event. The meaning compression principle "refers to the effect of the interaction of smaller-scale semiotic resources on higher levels where meaning is observed and interpreted. [. . .] The *meaning-compression principle* is a principle of economy [. . .]. In this way, a given combination of resources compresses, in its patterned arrangements, meanings which can be unpacked and integrated into a more specified semiotic configuration on a higher level of textual organization" (Baldry and Thibault 2006: 19).

In the Camfrog example, the environment needs to be interpreted by unpacking each underlying semiotic resource: layout, colour, written chat, video windows, nickname listings, and personal and customizable private chat windows. Systems of hierarchy are detectable in this spatial and logical organization (i.e., smaller-scale semiotic resources within higher levels of textual organization).

This preliminary discussion has served the purpose of outlining some basic multimodal properties of such a specific environment, but that are nonetheless present in other similar systems, to highlight how these properties are linguistically and functionally integrated, and how they may be co-deployed by participants in the meaning-making process.

We may ask how users interact in such an environment, how they co-deploy verbal, audio and visual resources to produce meaning and, more specifically, what linguistic strategies are called upon, and if and when mode-switching comes into play. To answer these and other questions, such as those related to gaze management, kinesics and proxemic patterns and use of spoken and written discourse, field observation was carried out over the course of about twenty-four months. The author played the role of one of the many lurkers, observing and being observed by other participants. However, because lurkers are as common as viewers in such environments, we may postulate that active viewers were not significantly disturbed by the author's silent presence. Along with online observation, several screencastings using Camtasia Studio 7 (© Techsmith Corporation) and Camstudio 2.0 (© Microsoft Windows) were recorded, to keep track of all the interactions and make possible subsequent repeated views and *ad hoc* annotations.

2.3.1   Digital Interactions and Ethics

By way of a preliminary remark to observations on our dataset, some reflections on ethical questions must be made at this point. Paul ten Have (1999: 61) reviews the research participants' rights to refuse: (1) to be recorded or to give access to the situation for recording purposes; (2) to grant permission to use the recording for research purposes; and (3) public display or publication of the recordings in one form or another. Borders between these three issues are often blurred, and researchers need to be aware that studies on video data require the establishment of trust between all stakeholders, because it involves extreme exposure of research participants. The latter may feel discomfort at different stages of the research project, even though they granted permission, as Yakura (2004) warns in her paper on "informed consent." Video recordings are, in fact, particularly delicate and sensitive data, as participants may be either wary or suspicious about being in a recording that is shown in a context (e.g., academic conference or publication) other than that of its original setting. Furthermore, Yakura discusses two different "gazes" (i.e., perspective from where a video needs to be viewed) that are especially salient in videotaping research: the "surveilling" gaze (a "public" perspective, for example video cameras at banks and convenience stores or the academic way to read and present video data) and the "reflexive" gaze (a "private" perspective, for example viewing ourselves in a home video).

In the US, issues about ethics are handled with the utmost care where human subjects are involved. Research in social sciences is regulated by the institutional review board (IRB), which examines how researchers deal with human subjects. In the case of web-based video observations, the question is rather intricate from an ethical point of view. Even though, technically, any user may be recording a video session (saving conversations is an integrated function in Camfrog), this open possibility does not equate with an automatic permission to record video interactions without explicit consent. We may also add that subjects may not be willing to be undergoing a "repeated, detailed examination" (Goodwin 1994), which any research observation requires. Banks (2001) challenges the researchers' right to *represent* others through different forms of representations, such as publications, lectures and so on, however, he rightly points out that "taken to extremes, all human sociality should cease, as all humans continually construct and disseminate representations of others" (Banks 2001: 129).

Participants have in this project granted permission to be studied and observed for research purposes, but their identities have been protected by my ethical choice to use drawings instead of screenshots, considering reservations about the publication of screenshots expressed by some informants.

Drawings present several difficulties: they are time-consuming and require specific expertise, so that they can be used selectively, only for very brief and fine-grained analyses. Furthermore, drawings incorporate the researcher and artist's bias that represent participants in their interactions. Conversely, screenshots are more objective and reduce the stakeholders' bias, but drawing preserves informants' identities, who could otherwise feel exposed if displaying their "real selves," despite informed consent. Moreover, drawings lack important information inscribed in screenshots and may also reflect the bias of analysts, who may also be unconsciously willing to interpret data in order to validate their assumptions or answer research questions. Factors affecting transcription may also include informants' feelings on the matter, technical issues or the analyst's research agenda. Comparing drawings with screenshot, we may argue that the former is representational and iconic, whereas the latter is mimetic and realistic. However, in the present research, drawings have been preferred to screenshots, primarily for ethical considerations, as research informants were "caught" during private, intimate and personal conversations, however paradoxical this may appear in an open, web-based environment, and despite informed consent. Furthermore, bias and representational issues are false problems, as *any reconstruction* of a live event is always biased and representational. To some extent, the same may be said of any kind of interaction that is *retold* and *re-presented* through different forms of narration.

These remarks on ethical issues are not limited to transcriptional issues that any researcher in the field of multimodal studies is involved in. They are also meant to raise awareness of the growing amount of intrusion, if not invasion, of personal/digital space on the part of several stakeholders, such as the marketing and advertising industry. Researchers in computer-mediated communication are particularly prone to overriding ethical considerations, due to an increasing blurring of personal and collective, private and public spheres occasioned by the unprecedented possibilities of intruding, indeed lurking, into the lives of others.

2.3.2 General Overview: Resources, Language and Participants

This chapter is focused on specific research questions, such as whether and why mode-switching is used and how it is distributed in spontaneous video-based interaction.

As a general overview of the data, it is crystal clear that in such environments, participants may choose from a wealth of semiotic resources, ranging from language (e.g., using mode-switch to alternate spoken and written exchanges and *vice versa*) and other resources (using resource-switching, that is alternating different resources across conversational threads, cf. Chapter 3). Resource selection is one among many choices for self-representation: for example, users select a nickname, which can also be altered in different communicative situations. According to their willingness to share personal information and to self-display onscreen, they can select portions of "items" (e.g., body, face, one's own room, office) as representative of the self.

With regard to the mode of writing, participants can choose their preferred colour, font, size or use of emoticons: all these are recognizable and meaning-making resources that accompany and ornate their identity, along with their, if any, projected image. Emoticons are frequently used, with an average rate of about ninety-two per hour (computed on fifteen randomly sampled hours of recording). Capitalization is used idiosyncratically: it can be ignored, used to highlight or focus on some specific item/concept (e.g., LOL, DON'T WORRY, I'LL PAY IT, DID YOU SEE THAT?) or, if a whole sentence is written in capital letters, to shout at someone (e.g., DON'T SAY THAT, THIS IS TOTALLY UNFAIR). The latter practice is not well received and implies a rising, provocative tone, even though prosody can only be *imagined* and *mimicked* through writing.

Punctuation is also idiosyncratic, mainly reflecting the desire to reproduce voice prosody on the part of users, but in a different way from that which is commonly found in conventional punctuation in formal writing. For example, a very common usage of commas is when they come after names or nicknames, signalling a pause. Furthermore, an occurrence such as: *Diamond,* indicates that the participant is directly referring to *Diamond_in_the_sky,* a moderator. It is very likely, in such occurrences, that the participant who uses this device wants to ask for the floor. The comma following a nickname or name may be occasionally substituted by suspension points. Users can also place a written caption under their projected video, to highlight a specific message that "surrounds" them for the whole time of the communicative event (e.g., "Happy Easter! ☺", "What to do on a boring day?" [*sic*]).

Such resources for writing are much more meaningful than in other settings, as is the case of one-to-one videochat, because they also play a seminal social role within the community, that is, that of establishing and making visible the community's hierarchies of power. These communities are embedded in contexts of situations marked by social roles, hierarchies, rules and duties. These rules and habits are not simply the basic "Terms of Service" established by the platform's owners: there are subtler rules of behaviour, less explicitly stated, but very rigorously applied, with punishments inflicted on those who do not comply with them. Every room has its own non-written rules. If users do not comply with them, they are reprimanded or banned from the community. Their microphone can be silenced or deactivated. In any case, a message appears on screen, announcing the punishment, as the following example shows:

> SpaceCowboy69 kicked dafernandem44 for **"No music or singing allowed in this room."**

Behaviour during interactions is monitored by users who hold special privileges, which are semiotically signalled by colour: red for room owner, green for moderators, blue for friends and black for simple users. Moderators and owners have the power and authority to reprimand other users and make them respect the rules:

> lemon_drop: you can't say that in here
> lemon_drop: You can't offend anyone in this room
> lemon_drop: Please read the room rules before you enter.

Interestingly, as soon as it comes to take on a formal role, users tend to adopt a more formal register, including the use of capital letters, where appropriate, as the example above shows. Akin to what will be contended in the following chapter, video communities are not as open as one might assume. These digital societies are rather exclusivist, and especially so when specific kinds of communities are involved. For example, the room with musicians banned me, even though I did not break any of the room's rules, on the grounds that I did not have the right to be in the community, because I am not a musician.

The mode of speech is equally complex, because it allows the use of audio, with ensuing transmission of voice and its related features, such as intonation, pitch and overall prosody. Using the microphone, participants also join an interactive conversational event that features many of the traditional conventions of face-to-face conversation, which nonetheless presents significant challenges, for example with reference to turn-taking and asking for the floor.

The choice of transmitting one's own video is not as obvious as it may appear at first sight. Different software programmes cater for different options, including the choice of *not sending* one's own projected image. However, as soon as participants decide to use this option, they are in the position of making a series of more or less conscious semiotic choices. First and foremost, they decide which self-representation will be performed: the subject may be the participant itself, but many other options are available, for example pictures, animated gifs (i.e., Graphics Interchange Format), settings (e.g., one's own room, office, kitchen, etc.), objects (e.g., tables, other laptops, wall with or without posters, or other decorations).

An important feature of such environments is the so-called *room*. The room is a virtual place, but it represents, for users, a tangible space, where interaction, conversation and other mundane events occur daily. The room is the background where all social actions are instantiated and the room is built around the idea that all semiotic resources, participants and their roles should be immediately and unequivocally recognized, even by novice participants. By way of example, it is similar to a religious event, where all items, symbols and equipment placed in the setting point to how and why things are the way they are. The same may be said about other highly ritualized events, such as court meetings and procedures.

Frequency and familiarity are also seminal to understanding how and to what effect participants are community members. Frequency of logging in, and time and degree of interaction are obvious criteria for user participation. Users can either visit the room occasionally and chat with semi-strangers or, conversely, be much more active and regular visitors, so as to create more sustained social bonds.

In the dataset, five kinds of participants have been identified: (1) users who visit the room daily, although not at a specific time of day or night. They may also leave their digital IT islands without logging out, leaving their video option on; (2) users who visit the room daily at around the same time, possibly increasing and changing time at weekends; (3) frequent but not daily users; (4) weekend users; and (5) occasional users.

Users belonging to the first group make up the social basis for the whole community, whereas those who belong to the second group tend to aggregate only with users belonging to the same group, as they visit the room at the same time of day (or night). Participants from the other groups are less easily "socializable," as it is not easy to catch up with more-regular users, and *vice versa*.

Frequency goes hand in hand with familiarity: frequent users have the chance to "meet" and get to know other frequent users, referring to each other using their proper names, instead of nicknames, hence establishing more intimate rapports with fellow participants, also apparent in the nature of communicative exchanges, marked by more informality and intimacy. Typically, frequent users are more likely to show their video and thus become familiar presences in the room, also for non-frequent or occasional participants. Non-reciprocity of acquaintance is very common, as the following example shows:

> dezzytheartist: have we spoken before?
> freemanoncam1: nope

dezzytheartist: or you just see me around so u recognized me?
dezzytheartist: loool
AndrewWythe: lol
freemanoncam1: i've seen u here

After being directly and familiarly addressed by *freemanoncam1*, *dezzytheartist* enquires as to whether they actually talked before, or if it is her popularity that makes her stand out.

Nicknames are self-appointed/chosen and are important both for the participant him/herself and for others. It is the label with which users prefer to be known, and, most importantly, it is the way in which others can recognize them. Nicknames are not eternal as real names are: we can change them and, thus, even modify our alter ego's virtual personality. Users often do change their nicknames, but according to their degree of popularity, other users are still able to recognize them. During the observations, popular users have changed their nicknames several times, but this has not hindered recognition, as the following example shows:

babygirlaivan: back
AndrewWythe: wb lemon
AndrewWythe: i mean babygiel
AndrewWythe: girl
babygirlaivan: LOL
AndrewWythe: sigh. . . .
AndrewWythe: sigh. . . .
babygirlaivan: get with it Andrew
AndrewWythe: sorry
babygirlaivan: lemon's not on duty :P
AndrewWythe: << such a noob
leveth: lemon
babygirlaivan: lol leveth
babygirlaivan: it's me
leveth: I know
leveth: lol

In the example, the popular user *lemon_drop* has changed her nickname to *babygirlaivan*, but is nonetheless recognized by other users.

What follows is a screencasting analysis of one transcribed and annotated multiparty video interaction (time: 22:09:28), which has been selected as a case study. Some shots and excerpts of spoken and written discourse are reproduced to shed light on some visual/kinetic and verbal features, respectively. Research participants are kept anonymous by deleting their faces and altering their nicknames. However, excerpts from text chat will be reproduced faithfully, by keeping the original font and size. Colour is not reproduced in this edition of the book.

2.3.3   The Development of a Transcription Model for Videochats

When analysing visual data, researchers have adopted different perspectives and lines of thinking (Prosser 1998; Baldry 2000; Emmison and Smith 2000; Hine 2000; Banks 2001; van Leeuwen and Jewitt 2001; Rose 2001; Norris 2004; O'Halloran 2004; Pink [2001] 2007). One aspect of my multimodal investigation is the *point* in the interaction at which users decide to mode-switch, which needs to be determined both quantitatively and qualitatively. Quantitative studies need to be focused on turn-taking management and mode-switching, measured through mode-switching frequency, duration of spoken and written turns (STs and WTs henceforth), and number of total and average mode-switches. Data are concerned with different interactional practices, namely those used in multiparty videochat conversation. These practices involve various numbers of participants and medium affordances. The ensuing diversity of strategies is examined accordingly. Data transcription, annotation and analysis are subsequently oriented and ultimately determined by medium affordances (e.g., differences between a multiparty videochat and a one-to-one videochat).

A major problem when analysing videochat conversations is transcribing and annotating them. As with face-to-face conversations, web-based multiparty videochats are not easy to transcribe, as they are chaotic in nature, for example featuring constant overlapping between conversational threads and modes. Furthermore, they are completely spontaneous and loosely monitored, unlike other video-based texts. For example, fictional texts creatively reproduce some portions of reality or manipulate the audience vision of reality. They are written, shot, produced and received as commodities or products for a specific goal (i.e., informative/referential, commercial or artistic texts of various kinds, such as films, advertisements, music videos, TV shows, etc.). Videochats are produced outside the economic chain of the buying and selling of products, and are not construed *a priori* as semiotic systems that fulfil a specific (i.e. commercial/artistic) goal. A videochat is not a text *per se*, but is a tool for interaction and *becomes a text* as soon as participants make it live.

Phases, subphases and transitions (Baldry 2004) are maybe less easy to identify, in that editing is, for obvious reasons, absent and a videochat looks like a long, uninterrupted single take (or tracking shot). Conventional gaze vector analysis, as summarized in section 2.1.4, is also impossible, for the reasons discussed above. As regards visual frames, the chief invariants are location and its features, which convey some information to a somewhat smaller extent, in that the image is always delimited and is typically fixed. What is in the foreground is a very close or close shot of both participants. Following Gibson's theory of perception (Gibson [1979] 1986), the delimited optic array "can specify visual kinaesthesis in the viewer even though the viewer may occupy a fixed position, seated, say, in a room with the screen occupying a specific position in front of him or her" (Baldry and Thibault 2006: 191).

However, visual kinesis takes on new connotations in a videochat, because this process is reciprocal and both participants find their visual kinesis realized in the other's "progressive picture" (Gibson [1979] 1986). The point of observation is, in fact, fixed and provides a static perspective for both participants. However, movements in videochats *can* be analysed and transcribed following criteria which describe configurations of semiotic relationships and define indexical orientation.

The dataset has been collected through *screencasting*, that is, video screen capture or digital recording of computer screen output, also containing audio narration. A screencast is a movie reproducing all events that appear on the monitor and, as such, seems to be a particularly flexible strategy when capturing dynamics (e.g., turns and moves) that come into play during a videochat, making them available to scientific scrutiny. This system has the advantage of encapsulating both events pertaining to videochats *per se*, and other parallel moves the participants are likely to perform: opening and closing other windows, surfing the net and so on, lowering their level of attention as regards their conversation at specific moments.

Screencasting is a useful method of recording, because it allows the exploration of other significant issues, such as the general low level of attention during a videochat, especially so when the context is informal and the other user has the floor. More specifically, multitasks performed during a video-conversation (e.g., checking emails or web pages) seem to demonstrate a parallel high level of reciprocal tolerance with regard to the low levels of the other participant's attention. This view may be partially explained by the consideration that a low level of monitoring of the discourse partner's gaze direction and impossibility of eye contact weakens reciprocal control. However, further research on multitasking and video communication is needed to test this assumption.

The screencasting method has allowed the creation of 300 hours of recording of web-based video interactions, which have been *qualitatively* analysed and subsequently transcribed and annotated. Multimodal approaches rely on transcription and annotation to make sense of all the data collected; however, these are never neutral and every decision taken with regard to transcription and annotation must be consistent with research rationale and agenda. Even though mode-switching is a fundamental area of investigation, other non-verbal semiotic resources, such as kinetic action, proxemic patterns and gaze management are taken into account.

Multiple participant interaction may be interpreted as a spontaneous conversation among a group of people, even though realized in a web-based environment. As in spontaneous talk, participants are mutually orienting to, and negotiating to achieve, successful, ordered and meaningful communication (Hutchby and Wooffitt [1998] 2008). However, the setting where this interaction takes place is multiple, in the sense that it is physically dislocated into as many places as there are participants. In other words, a multiparty conversation is *spatially fragmented and temporally consistent,* in that it is located in many different places (corresponding to each participant's physical collocation), but is tied together by the fact that it occurs at the same time.

As Flewitt et al. (2009) argue, transcriptions depend on research context and must be viewed as "reduced versions of observed reality where some details are prioritized and others are left out" (Flewitt et al. 2009: 45). The transcription presented in Appendix 2.1 is no exception. What are the priorities and how have they been highlighted? What elements are foregrounded or backgrounded? How was the transcription model developed?

As this volume's title suggests, spoken and written discourse are foregrounded, and especially so considering the way in which they alternate in digital media. As discussed in Chapter 1, speech and writing are blurred notions and this is particularly true in the context of spoken and written informal contexts vs. spoken and written formal contexts. To bypass this problem, Lakoff proposed a distinction between spontaneous and nonspontaneous discourse in the early Eighties (Lakoff 1982). However, digital systems of interaction, such as videochats, further blur this distinction when both modes are used at the same time and in the same interactional event. Thus, the notion of mode-switching becomes a central theoretical concept in this study and may be identified as a priority to be addressed in the following dataset analyses. A further consequence is that language will be in the foreground. Although other resources for meaning making have been discussed in this chapter as worthy of in-depth analysis where human communication is concerned, they are, nonetheless, placed in the background. This choice does not imply that they are deemed as secondary; they are simply less relevant to the chapter's concerns.

However, despite the setting up of research priorities, the development of a transcriptional model is neither strictly consequential nor straightforward (O'Halloran 2004). Flewitt et al. report on how different approaches have been identified and different units of analysis developed (2009). Turns of speech are classic units of analysis, but are problematical when other modes come into play, as is the case with dynamic multimodal environments. The case of videochats is particularly thorny in that they include both STs and WTs, which multiply the occasions for overlapping *across modes.* For these reasons, the temporal unit of seconds has been selected, following Baldry and Thibault's (2006) rationale. The temporal unit of analysis is particularly flexible in that it allows a linear reading, even though turns often overlap, also mixing speech and writing *within* conversational threads. In the example of transcription reported in Appendix 2.1 (henceforth A2.1), a matrix has been devised to tackle the videochat genre. It may be used to transcribe both multiparty videochats, as is the case in this section, but also one-to-one video conversations.

The first column, "Time," features the selected unit of analysis and organizes a temporally oriented, vertical reading of transcribed data. Conversely, a horizontal data reading "freezes" each time-span, reporting each participant's turn and other relevant data discussed below. Following the horizontal reading, the second column, "Participant," presents the name of the participant whose verbal (either spoken or written) exchange is transcribed. Gender is indicated (M) or (F), because users' nicknames do not always reveal gender information, which may be of interest when interpreting gender dynamics within videochat interactions, as shown in section 2.3.4.

In some cases, more than one participant is listed as belonging to the same time-span: this graphic device has been designed to signal that verbal exchanges occur within *that* time-span. In other words, when many participants are reported as speaking or writing in a single time-span, this means that speech and writing overlap and occur during that very time-span. The third column, "Speech," reports verbal transcription of STs. Transcription is highly stylized and does not engage with any kind of fine-grained analysis. The adopted conventions reproduce the casual nature of spoken conversation and are easily accessible to readers not familiar with phonological/prosodic symbols. Transcription conventions are reported in Table 2.4, while orthographic representation of fillers are listed in Table 2.5.

STs are never transcribed with an initial capital letter, as this is a carryover from written first word capitalization. The adopted spoken transcription

Table 2.4   Transcription conventions for spoken turns

Symbol	Meaning
//	completion (falling tone)
no end of turn punctuation	non-termination (no final intonation)
/	parcelling of talk, breathing time (silent beats in Halliday's 1985a/1994 system)
...	pause of ½ second or more
..	pause of less than ½ second
?	uncertainty (rising tone)
!	surprised intonation (rising-falling tone 5 in Halliday's 1994 system)
WORDS IN CAPITAL	emphasis, stress, increased volume
(...)	inaudible/untranscribable utterance
(words within parenthesis)	transcriber's guess
-	false start/restart

Adapted from Eggins and Slade ([1997] 2004) and Tannen (1989).

Table 2.5   Transcription conventions for fillers

uhm	doubt
ah	staller
mmm	agreement
eh	query
oh	reaction
ooh	surprise

Adapted from Eggins and Slade ([1997] 2004).

conventions follow a different rationale and, as such, need to be clearly distinguished from written orthography. Furthermore, each turn is allotted a separate slot in the matrix, so capitalization is unnecessary, if not misleading, for spoken discourse transcriptions. Only one size font is used to transcribe STs to differentiate them from the different fonts and sizes used by participants in WTs.

The fourth column, "Writing," reproduces the WTs produced by participants. Here, the transcription features a number of different fonts, sizes and colours, which faithfully reproduce what each participant used in the original interaction (note that colour is lost in this edition of the book). Whereas the "Speech" column is graphically poorer, but richer in prosodic descriptions, the column on writing is graphically richer, as it duplicates WTs in the way they were shaped by participants. The *graphic speech* notion, referred to in section 2.1.1, may be recalled here to highlight the dual nature of text chat in web-based contexts. The "Writing" column is accordingly less manipulated, while the "Speech" column is more open to the transcriber's interpretation.

The fifth column, "Mode-switching," is a seminal unit of analysis for the aims of this research. It is placed at the centre of the matrix and this is not casual. The temporal unit is in the foreground because it is placed on the left of our matrix and, as such, holds a privileged position in the reading of the whole transcription, at least in the context of Western visual literacy (Thibault 2000). However, the central position of the "Mode-switching" column reintegrates the relevance of the spoken/written alternation that is high on this volume's agenda. In this column, a minus (–) or plus (+) shows which turn is characterized by mode-switching. If a participant continues to use the same mode (either spoken or written) as that used in his/her previous turn, then a – signals that no mode-switch has taken place. Conversely, if a participant mode-switches, that is, changes the mode used in his or her previous turn, a + appears in the column.

The following column, "Posture," reports observations on body *positions* of visible participants. As this column is in some aspects similar to the following column, "Kinetic action," it has been decided to include descriptions with regard to static body positioning under "Posture" and descriptions on dynamic actions/moves under "Kinetic action." Posture is mainly focused on head and torso positions, on more or less *involuntary* movements and on more or less relaxed poses, such as standing or sitting. "Kinetic action," a definition taken from Baldry and Thibault 2006, includes *actions*, such as drinking, writing, smoking a cigarette. Despite the recognition that a subdivision between voluntary and involuntary movements/actions is controversial and a highly crude abstraction (e.g., is gesticulating voluntary or involuntary?), it is nonetheless useful to unpack the complex interaction that is going on in multi-room videochat client systems.

The eighth column, "Gaze," gives an account of observable gaze behaviour that is highly complex and difficult to track in videochats, for reasons discussed in section 2.1.4. Direct gaze, for example, is indicated as "looks at webcam," as this definition is closer to what happens from a technical point of view. Furthermore, as interactions in a multiparty videochat involve a large number of people, gaze is overall less focused than in one-to-one videochat conversations and, as such, gaze descriptions are rather vague.

The ninth and last column, "Staged proxemics," refers to the *representation* of distance for each participant. The term *staged* alludes to the notion of representation that may be more or less intentional on the part of participants: very close to medium shots of participants may be in effect due to technical constraints (e.g., webcam position) or to intentional strategies to *stage* distance between oneself and the rest of the (digital) world (e.g., webcam *positioning*), as discussed in section 2.1.2. The "Staged proxemics" column also includes some information relevant to the physical environment surrounding each participant. The projected environment is highly dependent on webcam positioning, in that very close shots do not allow for the inclusion of "environmental" elements, whereas medium shots include larger portions of the environment where every participant is *framed*.

Drawings are inserted in a further column, to facilitate reading and accompany written transcription in tune with a multimodal approach. Drawings are totally ingrained into the model from a theoretical point of view and complement written descriptions.

2.3.4   Observations on a Multiparty Video Interaction

The following observations are based on a multimodal corpus made up of 300 hours of recorded spontaneous web-based video interactions, drawn from a Camfrog room chat called *SpeakEnglish*. The field observation described here includes 10 to 14 visible participants and from four to eight lurkers (including myself) and lasts 22:09:28 minutes. It is partly transcribed in A2.1. The number of participants varies as users signed in and out during the recorded interaction. However, there were thirteen active users engaged in spoken and/or written conversation. Four subjects were the most active participants, three mainly using speech and one exclusively using writing and not broadcasting a video.

The main topics discussed by participants in the recorded Camfrog Conversation 1 (hereafter Cam1) deals with Camfrog's rules and affordances. Initially, *tod3344* is reprimanded by the moderator, *Diamond_in_the_sky* (hereafter *Diamond*), for his improper use of fonts in the written chat. The moderator only uses written chat, but two other experienced users, a man in his thirties, *essence*, and a woman aged forty-seven, *tinker64* (hereafter *tinker*), join in the discussion. Font size is a hotly debated issue and a high number of turns is devoted to defining which is best. *Tod3344* resents being scolded by the moderator and tries to make fun of him on a number of occasions. *Essence* contradicts the moderator's definition of the right size and a part of the transcribed conversation revolves around this issue, even though a general lack of coherence may be impressionistically detected in Cam1, especially so when taking consistency across spoken and written conversational threads into account.

The matrix reported in A2.1 transcribes 13:02:17 minutes and follows the criteria discussed. By way of a preliminary remark, we may note that following Cam1 is extremely difficult, if not impossible, even after several and repeated views. This may be said about *all* recorded interactions that constitute our corpus of analysis, and is due to the constant superimposition of STs and WTs. Observers (i.e., those who view the interaction *after* it has being recorded) should content themselves either with following the spoken *or* the written discourse. Appendix 2.2 reports speech transcriptions, following the same criteria as those followed in Appendix 2.1 (see Table 2.4). Giving priority to speech would prove equally frustrating, as Appendix 2.2 shows: without WTs, STs are barely comprehensible.

Conversely, the transcription presented in A2.1 allows a *vertical linear reading* of both STs and WTs, thus unpacking and often clarifying overlapped or superimposed turns. This linearization dissolves the double and superimposed nature of multiparty videochat interaction, but has the heuristic merit of disentangling spoken and written discourse via a zigzagged reading (from left to right and *vice versa* and from the second to the third column, and *vice versa, going downwards*).

The transcribed communicative event consists of one hundred turns. Quantitative information is reported below in Table 2.6 and gives a first insight into the number of words in both STs and WTs, their average length and also an indication of type and token word number in Cam1. Quantitative analysis was carried out using Wordsmith 6 (Scott 2012).

Qualitative observations were also gathered through digital field notes. As regards the content of discussions, which revolved around a wide range of topics, a high incidence of "metamessages" has been detected. At Georgetown University Round Table 2011, Tannen defined the "metamessage" as *a message on the medium*. Multiple-room videochats are a case in point. Lexical analysis reveals the presence of lexical items which can be

Table 2.6   Spoken and written turns in Cam1

Total number of turns	100
Spoken turns (ST)	45
Written turns (WT)	55
Total no. of word types per ST	299
Total no. of word tokens per ST	1010
Average number of words per ST	22.4
Total no. of word types per WT	197
Total no. of word tokens per WT	347
Average number of words per WT	6.3

grouped into a semantic area under the general heading of "medium related comments" or "metamessages," such as *mic, version* (referring to the software programme), *loud, low, font, size, rules, users, terms, Camfrog, hear, ice-ops, room, download, i-phone, app/application, warn, kick off, ban, IM, Kermit.*

Needless to say, no claims are made about any kind of statistical validity of the reported data, which are strictly limited to Cam1. STs are slightly more than three times longer than WTs. The average number of words per turn is overall low and the type and token word number suggests a number of hypotheses, for example as regards the significantly different type/token ratio in the spoken and written corpora. When it comes to analysing mode-switching, eleven mode-switches are detected across one hundred turns, 11 percent of mode-switching in Cam1. However, this apparently low level of mode-switching must be viewed in terms of the relative time frequency in Cam1: slightly less than one per minute.

Essence is the highest *mode-switcher,* with seven mode-switches in Cam1, while *tinker* and *Sensei* each mode-switched twice. However, in the following 09:07 minutes of that recorded interaction, four more mode-switches took place, two by *Sensei* (speech-to-writing and writing-to-speech), one by *tinker* (speech-to-writing) and one by *Royal* (writing-to-speech). Furthermore, three long WTs substantially increase the number of words per WT. A young man nicknamed *Vancouver* tells a splatter story in three long WTs, respectively made up of 83, 96 and 70 words, for a total number of 249 words, thus almost doubling the number of words in WTs produced until *Vancouver* comes in. Despite the high number of swear words and detailed and disgusting descriptions of corporeal functions and illegal drug trading in which *Vancouver* was involved, the moderator does not warn *Vancouver,* nor does anyone comment on or criticize his improper use of language. Quite strikingly, font size appears to be a more serious matter than using improper language or even hinting at illegal activities.

Before going into further detail with regard to Cam1, let us now introduce the heuristic model into which data can be fitted. This model has been devised *a posteriori* for an in-depth examination of all the video data making up the corpus. It is presented *before* a full discussion of Cam1 data only to facilitate data reading and suggest possible interpretations.

2.3.5   Reflections on Mode-switching and Data Discussion

Overall interaction may be described as occurring on at least two higher-level organizations coupled to the two modes of speech and writing. Participants are thus defined on the basis of their preferred mode of interaction *exclusively for the purpose of this specific analysis* (even though this rather crude definition may be used in other analyses for practical reasons) and divided accordingly into (1) users talking and asking for the floor by clicking on the "Talk" icon (henceforth, *speakers*) and (2) users communicating exclusively via the written chat (henceforth, *writers*). The subdivision of participants into the group of speakers and writers is devised only for descriptive purposes.

As anticipated, both speakers and writers generally carry the interaction forward without mode-switching. This was observed not only in Cam1, but in the *whole* video corpus and tested against the backdrop of several hundred hours of video interactions. To put it simply, those who talked did not write, and those who wrote did not talk. Turn-taking adhered to each mode and was technologically constrained (i.e., speakers held the floor by clicking on the "Talk" icon). The maintenance of this interactional status was observed several times during all field observations and suggests a low level of mode-switching. However, several "breakings" of these general levels of organization have been noted in the corpus-based multimodal analysis carried out at both the macro-level of analysis (the whole corpus of video data) and micro-level (Cam1). The *average* mode-switching ratio amounts to about 16 percent in the general corpus and, therefore, needs to be addressed as a research issue worthy of further investigation, also to a great extent matching data that emerged from the survey discussed in the second section of this chapter.

Continuing with our micro-level analysis, it may be argued that speakers were the most active users, both in terms of number of turns and general continuity and coherence with regard to the main topics discussed. Despite the slightly higher number of WTs, writers were less engaged in the main thread, but interspersed this with non-related subthreads (mainly greetings to other users, or comments completely unrelated to any conversational thread, such as turn 01:28:23 by *kentish*). Speakers irregularly interacted with writers by using written chat, thus mode-switching. No writers mode-switched during Cam1. A summary of these findings is shown in Table 2.7.

Table 2.7   Mode-switching in Cam1

Kind of participant	Number	Mode-switch	Number of mode-switchers	Number of mode-switches
Speakers	3	yes	3	11
Writers	10	no	0	0

All speakers therefore mode-switched in Cam1, while no writer mode-switched. We may assume that writers did not have the chance to mode-switch, as they were not able to resort to video conversation for some reason or another. For example, they were not equipped with a webcam or a microphone. In this sense, mode-switching seems to be constrained by technicalities. However, why did speakers resort to writing when they could simply have spoken? A preliminary hypothesis was that users, when in possession of *full* medium affordances (i.e., they could both speak and write), tended to answer in the same mode as that used by the questioner. In other words, if addressed by a speaker, they preferred to answer by speaking, and if addressed by a writer, they tended to write their answers.

To account for mode-switching rationale, we may draw some useful insights from conversation analysis. The latter deploys different methods of analysis and is concerned with different research goals, but what it shares with this research project is the interest in *mundane and spontaneous talk-in-interaction* (cf. Eggins and Slade [1997] 2004; Hutchby and Wooffitt [1998] 2008), which is also reflected in the rationale of speech transcription, which has been used here by adapting conversation analysis transcription conventions.

By the same token, *mode-switching may be interpreted as a general way of managing turn-taking and, to be more specific, to avoid or bypass trouble caused by technical problems deriving from the medium, to repair misunderstandings or to increase opportunities for asking for the floor without being impolite.* In effect, overlapping across modes is not only perfectly acceptable within the multiparty videochat genre, it is also encouraged by some participants.

Mode-switching can thus be described as:

1. *Self-initiated* (i.e., when Participant A mode-switches on his/her own initiative, which may be more or less overt, for example to ask for a secondary floor when the other floor [both spoken and written] is held by Participant B);
2. *Other-initiated* (i.e., when Participant A mode-switches being *directly or indirectly* prompted by Participant B, who addresses A in a mode different from that used by A).

Empirical examples are drawn from the small sample collected in A2.1. For *self-initiated mode-switching*, several examples are found in *tinker*, who mode-switched on a number of occasions when the spoken floor was held by someone else. Another example of *self-initiated mode-switching* is found when *essence* resorts to writing at 07:59:22 to make fun of *tinker*, who was previously asked for her marital status by *Diamond*. For *other-initiated mode-switching*, *tod3344* asks "are my fonts the right size now?" using text chat at 00:38:12. He is purposely using a very small font size (almost unreadable) to contest the moderator's reprimands. The irony is not lost to *essence*, who replies through mode-switching at 1:09:04. In this example, *tod3344* is not *directly* addressing *essence*, who nonetheless joins in to sort out a possible conflict. *Essence* thus uses the same mode used by *tod3344*, even though he is not directly called to action by *tod3344*.

Another significant use of mode-switching is, as anticipated, a way to make sense of transition relevance places. In a multiparty videochat system, a large number of users talk at the same time. Technical problems, such as bad audio quality (discussed from 03:45:11 to 05:27:09 in Cam1) or delay in the transmission of video data, may pose problems in projecting transition relevance places for turn-taking. The problem is partially solved if we consider that, as already discussed, no overlap is possible among STs, but overlap is very frequent *across modes*. Participants are, thus, in the problematical situation of managing turns when mode selection is open to anyone, making projections more troublesome. These problems are apparently avoided or bypassed, thanks to the maintenance of *multiple floors* for conversations. Mode-switching thus becomes an essential linguistic strategy to bid for cooperation and to keep this desired state of affairs without being seen as impolite owing to repeated misunderstandings of transition relevance places.

Repair (Schegloff, Jefferson and Sacks 1977) is also made possible by mode-switching. At 04:20:04, *tinker* writes: "Sensei your audio's fine" using the text chat. After she checks that *Sensei* has not replied after three turns, she mode-switches and repeats the message, making explicit reference to a possible misunderstanding: yeah /and sensei /if you didn't understand that/ sensei your audio's fine//. However, this happens rarely and mode-switching is used for repair on a number of special occasions (for example, after repeated and failed attempts at making oneself heard/read). Normally, users repair in the preferred or selected mode of interaction. By way of example, at 02:25:12, *tinker* says: good morning sensei/ and goodnight tea/ sleep well sweetheart//. *Essence* replies: hey did you say essence? no/ no/ no/ I know/ I know/ I wake up like three hours ago//. *Essence* thus performs a self-initiated repair (Schegloff, Jefferson and Sacks 1977) without mode-switching. Another example may be found at 06:16:08, when *Diamond* writes: "Essence its 6.090 the new version." *Essence* thus replies: oh . . then it's nine/ sorry/ sorry/ it's eight//. This is an example of *other-initiated other-repair* (Schegloff, Jefferson and Sacks 1977) performed by *Diamond*, who does not mode-switch.

As regards other information collected and transcribed in A2.1, such as posture, kinetic action, gaze and staged proxemics, the short interaction does not allow for generalizations. However, some non-verbal data are self-evident and may be interpreted using sound common sense: when *tinker* is directly addressed by *Diamond* at 07:38:11: "You married by any chance?" her posture (she pushes herself upright), kinetic action (smiles) and gaze (more focused) openly reveal her interest, as documented in A2.1. After *Diamond* writes: "I'm engaged lol ☺", her non-verbal language abruptly changes, showing disappointment, as is also attested in her two subsequent STs at 08:30:07: yeah/ I think engaged people should wear SIGNS/ like toilets do// and at 09:04:13: have you ever seen/ an engaged sign on the toilet door? I think all engaged people should wear signs saying/ ENGAGED//.

The model presented allows the collection of video data, linearization of micro-level analyses and a complete replicability of application to other web-based video/text environments. However, a limitation may be found in the extremely thick description that is considerably time-consuming for analysts. A thorough qualitative analysis is needed on larger samples of video data to facilitate the subsequent selection of small data sets.

Another limit is constituted by the fact that speakers are described much more in their non-verbal language use than are writers in Cam1 (but also in other similar contexts). This is due to the fact that *most* writers do not broadcast a video of themselves in multiparty videochat client systems. This limit cannot be solved at the moment, even though many interesting observations may be found in WTs. Each writer exhibits different concerns according to their use of visual resources. We may note how *Diamond* is careful to avoid the typical sloppiness of text chatting, by several attempts at being orthographically correct and also paying attention to punctuation: "'size 12' is fine" (using inverted commas); "may I remind you once all again this is a [sic] 'ALL AGES ROOM' please mind the language" (cf. use of inverted commas, capitalization, formal register, but grammar mistakes, e.g., "a" instead of "an"); "essence but if you have a 'big' font, users will think you're shouting at them or trying to get a point across" (cf. use of inverted commas and explanation of font use rules in the community).

By way of conclusion, both quantitative and qualitative observations carried out on the *whole* multimodal corpus may be summed up in the following terms:

- Interaction may be divided into two main conversational threads: the spoken and the written.
- Both speakers and writers mode-switched on a number of occasions.
- Speakers' mode-switching exhibits higher incidence and frequency.
- Speakers use mode-switching as a general way to manage turn-taking, to avoid or bypass trouble caused by technical problems, to repair misunderstandings or to increase the opportunities for asking for a secondary floor.
- Writers are usually those who did not broadcast a video of themselves.

- Speakers are the most active participants and those involved in the highest number of verbal (both spoken and written) exchanges.
- Being the most active participants, speakers use written chat when they do not want to wait for the speaking floor, by asking for a "secondary" floor, that of written discourse.
- Writers are usually less engaged in verbal exchanges and show a consistent tendency to drop the main topic thread in favour of a (written) sub-topic.
- Writers overall are less eager to keep topic coherence than speakers.

This discussion has been based on A2.1, but final observations are based on the overall corpus. Several other observations may be made using the transcribed data, even though the time span is very limited. The aim was merely to illustrate method and research priorities. This transcriptional method can be read using a wealth of different interpretative models, such as psycholinguistics, ethnography, conversation analysis, discourse analysis, critical discourse analysis, pragmatics and corpus linguistics. Even though the present research draws *some* of its tools from *some* of the above-mentioned disciplines, it is based on the very general theoretical underpinnings of multimodal corpus linguistics, intended as a general and flexible *research method* based on multimodal data sets (e.g., audio/video/visual). Such a new paradigm of analysis opens the way for an arguably complex arena of discussion.

2.4   FETISH OR *SIMULACRUM?*

The reification of social relationships brought about by information and communication technologies entails a "fetish" of representations, something akin to what Marx called "commodity fetishism" ([1867] 1976). In his *Capital,* Marx argued that it is via commodity consumption that people experience social relations ([1867] 1976). Images then gain status and gradually substitute social relations in Marxist and post-Marxist theories (Althusser 1971). The media sociologist Baudrillard takes up this argument by saying that the rise of a culture of images produces a crisis in representation itself ([1981] 1994). Representation thus becomes more real than what is actually represented. Authentic, privileged or original meanings have vanished in contemporary society and what remains are *simulacra.* Baudrillard developed the concept of *simulacra,* claiming that what we consume from media becomes more real than what it purportedly refers to. He distinguishes four phases in image representation where image: (1) is the reflection of a basic reality; (2) masks and distorts a basic reality; (3) masks the absence of a basic reality; and (4) bears no relation to reality whatsoever; it is, as such, a *simulacrum* ([1981] 1994).

In investigating web-related texts, and the videochat in particular, we may ponder what kind of relationship is established between the projected image and its referent. Is the projected image a reflection of a basic reality? Is there a stable connection between the individual and his/her projected image? Or does the image mask and distort a basic reality, in ways discussed in this chapter?

Finally, we may speculate as to whether self-projections are a mere *simulacrum* of the individual, bearing no meaningful relation to the referent, that is, the individual. Discourses about the role of body and its place, entitlement and affordances in a virtual/real world have addressed questions about the body's presumed "disappearance" and on its (either feared or invoked) replacement by technological prostheses (Stelarc 1991). However, while some theorists of body dematerialization believe that "the body is obsolete" (Stelarc 1991), De Kerckhove points out that "this is inverted romanticism, very far from the underlying psychology of our incipient technological symbiosis. Most electronic technologies are not leading to the abandonment of the body, but to the remapping of our sensory life to accommodate a combination of a private and collective mind" (De Kerckhove 1997: 187). The aim of the research into videochat is precisely to make a contribution to this latter position, viewing the videochat as an example of the evolution of the human body *in toto* in relationship to man's natural environment, technology and communication in general.

Theoretical expectations, users' self-perceptions and descriptive models deriving from "traditional" multimodal criteria fail to account for all the facets implied in the multisemiotic complexities of videochats. The three-layered organization of the chapter has been devised to tackle this complexity. In the first layer, theories and models have been reviewed with the constant effort to slot together similar and contrasting theoretical components. A second layer has tested these expectations against the backdrop of observations from users who inhabit this digital world and, as native inhabitants, have the right to speak for their own world. Finally, a third step has been taken to carry out a direct observation of this world and its digital inhabitants, to further complement this project's theoretical foundations with empirical validation.

Despite the somewhat problematic investigative issues relating to transcription and annotation, together with possible contamination of data, the study of videochats and mode-switching appears to be a worthy enterprise. However, some implicit problems deriving from a page-based transcription must be addressed with the aim to overcome linear and page-oriented systems of transcription and annotation (O'Halloran et al. 2010).

Participants do not ultimately interact with each other in videochats, but with the medium. The context of situation is both shared and *not* shared at the same time. Each participant is an active producer of illusions for the other, in orienting his/her gaze, in exploiting verbal and non-verbal strategies to create a fake (or virtual) environment for the other, which is only partially real. Ultimately, the backbone of reality is constituted by interaction with a machine. Verbal and non-verbal strategies add to a sense of reality while dissolving it at the same time.

Appendix 2.1
Videochat Transcription

Participant	Speech	Writing	Mode-switching
Diamond (M)		but essence you're a regular so please follow the rules ☺	–
essence (M)	yeah/ I didn't say nothing/ ah.. improper or .. on his room/ tell me what I break/ because that/ I - I copy the rules/ yeah/ makes the rule/ that's all//		–
tod3344 (M)		ARE MY FONTS THE RIGHT SIZE NOW?	–
essence (M)	yes tod/ tod you can make/ it SMALL you know/make it bigger/ like on the 16/ 14 something like that/ used for it like this/ y'know//		–
Diamond (M)		Tod3344, Please Respect all users.	–
essence (M)		*they<small... but u change on red*	+
kentish (F)		SpeakEn	–
essence (M)		*use 14 or 16*	–
Diamond (M)		"size 12" is fine	–

sture	Kinetic action	Gaze	Staged proxemics	Drawings
ght hand on right temple, lateral leaning	Small head movements, touches faces with left hand	Oblique gaze	Very close shot, dark setting, no object visible	
ght hand on right temple, lateral leaning	Small head movements, turns head	Direct gaze, looks at webcam	Very close shot, dark setting, no object visible	
ily head visible, oblique position	Writes	Looks at keyboard then oblique gaze	Very close shot, dark setting, no object visible	
ily head visible, oblique position	Stands still	Looks at webcam	Very close shot, dark setting, no object visible	

(continued)

Participant	Speech	Writing	Mode-switching
essence (M)	well/ you know what? what I like if they use like.. uhm.. different/ because if you like chat with someone/ ah.. you know/ you're gonna click an action/ or make see this color or something/ we (wait here)/ I love you by the way/ ah.. and you just synchrony/ you see this color/ and it's all/ then you know/ that this person likes something in the room//		+
Tock66666 (U)		☺ ☺	–
Tea (F)		**goodnight**	–
Tock66666 (U)		**G N**	–
Tod3344 (M)		☉⊱◆°❨☉⌒↦ ∘⌒⊰⟨ℸ⌐ ᵒ⌐⊰↦	–
Sensei	so (. . .) I got this new update/ whatever/ that's cool//		
Tinker (F)	good morning Sensei/ and goodnight Tea/ sleep well sweetheart//		–
essence (M)	hey did you say essence? No/ no/ no/ I know I know/ I wake up like three hours ago/ I don't know… and you heard that she woke up/ she this mornin'		–

...ture	Kinetic action	Gaze	Staged proxemics	Drawings
...ight hand on right temple, lateral leaning	Moves left hand, gesticulating Points with finger Moves backwards Drinks from a bottle	Looks down No gaze discernible	Very close shot, dark setting, no object visible Plastic bottle enters frame	
...rontal position, head slightly bent forward (to keyboard)	Stands still	Looks slightly down	Medium shot. Personal room, posters in the background	
...ateral position due to webcam positioning, sits at office desk	Arms on lap, stands still	Looks in front of her (perceived by viewers as lateral gaze due to webcam positioning)	Medium lateral shot. Desk, well-illuminated room with Oriental ideograms on posters on the wall	
...ight hand on right temple	Slightly moves head backwards	Looks slightly downwards	Very close shot, dark setting, no object visible	

(continued)

Participant	Speech	Writing	Mode-switching
Diamond (M)		yes, essence but if you have a "big" font, users will think you're shouting at them or trying to get a point across	–
Diamond (M)		across*	–
Diamond (M)		so an bigger then 12 is to big ☺	–
King (M)	and you?/ you hear me?		–
essence (M)	well not really/ you have like/ uhm.. maximum 16/ you know/ so in day like/ uhm/ they give you a chance to use like 16/ and you can use it/ uhm .. uhm		–
Diamond (M)		please re-read	–
Gyss (U)		**HI ANGEL**	–
Diamond (M)		essence but if you have a "big" font, users will think you're shouting at them or trying to get a point across.	–
Sensei (M)	I just gauged my ear//		–
Tinker (F)	Sensei /at least you can talk this morning/ you don't have to hang on to the mic//		–

...ture	Kinetic action	Gaze	Staged proxemics	Drawings
...ight hand on right temple ...oves left hand backwards and forwards	Stands still, gesticulates with left hand spread open Shrugs	Looks slightly downwards Looks at webcam	Very close shot, dark setting, no object visible	
...rontal position, head slightly bent forward (to keyboard). Naked torso visible	Stands still	Looks downwards	Medium shot. Personal room, posters in the background	
...ateral position due to webcam positioning, sits at office desk	Arms crossed in her lap, Rocks slightly in chair	Looks in front of her (perceived by viewers as lateral gaze due to webcam positioning)	Medium lateral shot. Desk, illuminated room. Oriental ideograms on posters visible on wall	

(continued)

Participant	Speech	Writing	Mode-switching
Sensei (M)	I'm all right/ like it's supposed to be//		–
Tinker (F)	yeah/ it's all cool/ calm and collected/ this morning/ thank God//		–
Sensei (M)	it's.. it's my mic really loud/ like cause I - I just downloaded this/ like the new version or whatever/ and it's/ my mic it's saying that's like going all the way around/ like it's really loud or something//		–
Essence (M)	the (reaction) is good/ if it is called like.. like maximum/ you know/ I dunno (…) like ah I can't connect if it's like red/ then and it is like uhm/ it is like all the time/ you know or .. I don't know how to explain it/ but yeah/ IT is on the red and you're hooked//		–
doni_ada (F)		Jk	–
Doni_ada (F)		miss tinker	–
XSANDx (F)		**good mic on my end**	–
Diamond (M)		Sensei you using 6.090 of Camfrog? ☺	–
XSANDx (F)		**audio is perfect**	–

sture	Kinetic action	Gaze	Staged proxemics	Drawings
ontal position, head slightly bent forward (to keyboard). Stands still, naked torso visible	Moves head slightly	Looks down slightly	Medium shot. Personal room, posters in the background	
ateral position due to webcam positioning, sits at office desk	Half-smiles, rocks slightly in her chair	Looks in front of her (perceived by viewers as lateral gaze due to webcam positioning)	Medium lateral shot. Desk, illuminated room. Oriental ideograms on posters visible on wall	
	Stands still Rubs hand on chin	Looks to the left and right Frowns	Medium shot. Personal room, posters in the background	
ight hand on right temple	Gesticulates with left hand Moves backward and forward slightly Stands up	Looks slightly downwards	Very close shot, dark setting, no object visible Exits from frame	

(continued)

Participant	Speech	Writing	Mode-switching
Tinker (F)		*Sensei your audios fine*	+
tod3344 (M)		SOUNDS GOOD SENSEI ... KEEP UR EYE ON THE CAMFROG POLICE ☺	–
Sensei (M)		**yeah im using the newest version its all different lol**	+
Sensei (M)		**bullet**	–
Tinker (F)	yeah/and sensei/ if you didn't understand THAT/ sensei your audio's fine//		+
Tinker (F)	sensei/ I'm sticking with five point five/ until diamond - until he stop usin'it/ then I'll upgrade//		–
tod3344 (M)		I GOT MY ASS KICKED FOR USING THE ILLEGAL FONT	–

›ture	Kinetic action	Gaze	Staged proxemics	Drawings
›teral position due to webcam positioning, sits at office desk		Looks in front of her (perceived by viewers as lateral gaze due to webcam positioning)	Medium shot. Desk, illuminated room. Oriental ideograms on posters visible on wall	
			Medium shot. Personal room, posters in the background	
			Medium shot. Personal room, posters in the background	
›teral position due to webcam positioning, sits at office desk	Rocks slightly in chair	Looks in front of her (perceived by viewers as lateral gaze due to webcam positioning)	Medium shot. Desk, illuminated room. Oriental ideograms on posters visible on wall	
›teral position due to webcam positioning, sits at office desk	Rubs chin	Looks in front of her (perceived by viewers as lateral gaze due to webcam positioning)	Medium shot. Desk, illuminated room. Oriental ideograms on posters visible on wall	

(continued)

Participant	Speech	Writing	Mode-switching
Sensei	uhm .. I see/ uhm.. yes/ pretty cool/ like uhm .. I can show you what it looks like on my cam/ if you wan'it / it's like totally different lookin'//		+
Diamond (M)		Tod3344, they are no camfrog police, people that warn you are called: Operator's or Owner's.	
XSANDx		*I didn't get the new version yet...*	–
Essence (M)	it's like a six zero (nine)/ now you have six zero EIGHT/ and by six zero eight it's different/ than .. you can see a cam if there're pro users and/ you know/ if you go on the user/ and nickname like camera/ status/ watching/ age/and ok		–
Diamond (M)		I've seen the new version jelens_bre showed me ☺	–
Diamond (M)		Essence its 6.090 the new version	–
essence (M)	oh then it's nine / yeah/ sorry/ sorry/ it's EIGHT//		–
Diamond (M)		an there's also a 6-1 version of ice-ops	–
essence (M)	I use Kermit/ I don't know/ I got used to the Kermit//		–

sture	Kinetic action	Gaze	Staged proxemics	Drawings
rontal position, head slightly bent forward (to keyboard). Naked torso visible	Rubs chin	Looks left and right as if scan-reading some text	Medium shot. Personal room, posters in the background	
ight hand on right temple Moves left hand with cigarette backwards and forwards	Smokes a cigarette		Very close shot, dark setting, no object visible	
ight hand on right temple, slightly moves backward, then turns head to right	–	Looks to the left and right	Very close shot, dark setting, no object visible	
ight hand on right temple, moving slightly	Smokes	Lateral gaze, looks left and right	Very close shot, dark setting, no object visible	

(continued)

Participant	Speech	Writing	Mode-switching
Diamond (M)		Tinker I have a question for you ok? ☺	–
essence (M)	I have a feeling somebody's watching//		–
Tinker (F)	yeah / go ahead diamond/ what's - what's the problem?		–
Diamond (M)		You married by any chance?	–
Tinker (F)	no/ I'm not/ I'm single//		–
Sensei	do you see it? how it's different/ like you were on the (…) version is (…)/ is on my mic is like going up/ like in a circle/ or whatever/ that's what I was asking if it was too loud or not//		–
Diamond (M)		ah ok	–
Diamond (M)		who here is married?	–
Essence (M)		*Tinker* ☺	+
Essence (M)	no/ it's not too loud/ because it's not on the red//		+

sture	Kinetic action	Gaze	Staged proxemics	Drawings
ight hand on right temple, moving slightly	–	Lateral gaze	Very close shot, dark setting, no object visible	
ateral position, due to webcam positioning. Hand in front of mouth	Pushes herself upright. Stands upright, looks more interested	Eyes more focused	Medium shot. Desk, illuminated room. Oriental ideograms on posters visible on wall	
			Very close shot, dark setting, no object visible	
eeps standing upright, slightly moves to left, thus having a more central position	Smiles, right hand under chin	Looks downward	Medium shot. Desk, illuminated room. Oriental ideograms on posters visible on wall	
A	NA	NA	NA	
A	NA	NA	NA	
A	NA	NA	NA	

(continued)

Participant	Speech	Writing	Mode-switching
Essence (M)	it's perfect/ yeah//		–
Tinker (F)	I have no idea/ who is married here/ and who's not married diamond/ I - I just know/ that I'm not//		–
Diamond (M)		I'm engaged lol ☺	–
Tinker (F)	yeah/ I think engaged people should wear SIGNS/ like toilets do//		–
XSANDx		**With the new version can I block people from viewing my webcam?**	–
Essence (M)	no/ you cannot block that/ somebody cannot view/ right.. uhm		–
Diamond (M)		oi i'm saving for the ring my Fiance wants, thank you very much	–
XSANDx		**K**	–
Tinker (F)	have you ever seen/ an engaged sign on the toilet door? I think all engaged people should wear signs saying/ ENGAGED//		–

...ture	Kinetic action	Gaze	Staged proxemics	Drawings
..A	NA	NA	NA	
...ateral position, due to webcam positioning	Smiles, left hand under chin	Looks in front of her, slightly downwards. Very close to a direct gaze	Medium shot. Desk, illuminated room. Oriental ideograms on posters visible on wall	
...ateral position, due to webcam positioning	Stops smiling left hand under chin. Half-smiles	Looks in front of her, frowns	Medium shot. Desk, illuminated room. Oriental ideograms on posters visible on the wall	
...eans to the right. Right hand on right temple, moving slightly	Gesticulates with left hand	Looks left and right, then a quick direct gaze	Very close shot, dark setting, no object visible	
...ateral position, due to webcam positioning	Rocks slightly in chair	Looks in front of her (perceived by viewers as lateral gaze due to webcam positioning)	Medium shot. Desk, illuminated room. Oriental ideograms on posters visible on wall	

(continued)

Participant	Speech	Writing	Mode-switching
Diamond (M)		xSANDx, nope you can only "Pause"	–
Diamond (M)		haha ahh I see tinker ☺	–
XSANDx		**Ty**	–
Sensei (M)	not married /thank God		–
Diamond (M)		**Yw**	–
Tinker (F)	yeah/ ah/ Sensei/ have you tried the new/ ah.. the new ice-ops six one?		–
Sensei	uhm .. the ice-op .. I don't know what that is//		–
Tinker (F)	yeah/ it's the operator's tool pack that goes with the/ ah.. Camfrog//		–
Sensei (M)	oh you mean like Kermit .. or whatever/ that what you're talking about/ I-I never like figured out how to use it/ cause I came in an open room one time/ they told me to download it/ uhm/ and I downloaded it/ I ain't trying usin'it/ I just couldn't figure it out/so ah .. I just did it from/ this .. just Camfrog//		–

sture	Kinetic action	Gaze	Staged proxemics	Drawings
rontal position, head slightly bent forward (to keyboard), hand under chin	Jiggles. Pushes head backwards	Looks slightly upwards	Medium shot. Personal room, posters in the background	
ateral position, due to webcam positioning. Left finger on her lips	Rocks slightly in chair	Looks in front of her then to her left	Medium shot. Desk, illuminated room. Oriental ideograms on posters visible on wall	
rontal position ans to left, lips as in an interroga-tive, dubious fashion	Left hand on chin	Questioning glance	Medium shot. Personal room, posters in the background	
JA	NA	NA	NA	
rontal position, head slightly bent forward (to keyboard). hrugs	Gesticulates with right hand	Frowns Looks up. Looks in front of him Frowns	Medium shot. Personal room, posters in the background	

(continued)

Participant	Speech	Writing	Mode-switching
Essence (M)		*yea*	+
Essence (M)	I don't know/ you've a (choice of many cams) / if you wish//		+
Tinker (F)	yeah sensei/ah .. ice-ops are very simple to use/ like Kermit's very simple/ ah.. once you get the hang of it//		–
Essence (M)		*yea*	+
Sensei (M)	eh that's ok/ ah/ I don't know /if I'm using it anymore anyway/I'm not in an open room anymore so.. cause there's too much RESPONSIBILITY/ you know/ I'm just here to chat/ you know/ talk to people//		–
Diamond (M)		sensei, don't worry you don't need ice-op really im a op for 3 rooms an ice-ops basically for a quick ban/ kick a bloc mics	–
Diamond (M)		ainly used for "setting topic"	–
Royal (M)		what no brain cells in the room yet.	–

sture	Kinetic action	Gaze	Staged proxemics	Drawings
			Very close shot, dark setting, no object visible	
NA	NA	NA	NA	
ateral position, due to webcam positioning	Rocks slightly in chair	Looks in front of her	Medium shot. Desk, illuminated room. Oriental ideograms on posters visible on wall	
eaning to the right. Right hand on right temple, moving slightly	Writes	Looks down	Very close shot, dark setting, no object visible	
rontal position, head slightly bent forward (over keyboard)	Hand on chin. Quick hand stroke	Frowns, looks downwards	Medium shot. Personal room, posters in the background	

(continued)

Participant	Speech	Writing	Mode-switching
Essence (M)	(…) a great op and fast/ and… this not the talk/ you just simply/ ah.. write/ and it's open/ you know/ just click on the right (goof) and it's copied to correct/ his name-nickname/ and you have like a warn/ warn if you wanna warn them/ it is kick/and he kicked up/ and you can make choices like/ what you gonna use like/ ah/ the face on the cam or/ I know you like (…)		+
Diamond (M)		*mainly	−
Royal (M)		Thear one now	−
Essence (M)	I mean/ all you gonna warn though/		−
Diamond (M)		Ice–Op is mainly for setting the Room Topic	−
Tod3344 (M)		I JUST FARTED ON MY MIC … ANYBODY HERE IT? ☺	−
Essence (M)	yes/ I did/ Tod/ oh my God/ that's nasty man/ how can you be like that?		−

ture	Kinetic action	Gaze	Staged proxemics	Drawings
A	Coughs	NA	NA	
A	NA	NA	NA	
A	NA	NA	NA	

(continued)

Participant	Speech	Writing	Mode-switching
Sensei (M)	I can smell it through my speakers (laughs)/ oh yeah/ like I said Tod/ you know/ I really don't care/ if I have that ice-op or Kermit or whatever/ you know/ just keep it simple and basic//		−
Essence (M)	and didn't he say his volume on or something/ and that'd be interesting/ ah.. for the coarse/ like he's gonna fart//		−
Diamond (M)		Excuse me, may I remind you once all again this is a "ALL AGES ROOM" please mind the language	−
Sensei (M)	did I say something bad? Because sometimes I - I curse/ without even knowing that I do it/ because that's just how I talk/ and.. yeah .. sorry if I did that//		−
Essence (M)	well /no/you didn't//		−

sture	Kinetic action	Gaze	Staged proxemics	Drawings
contal position, head slightly bent forward (over keyboard)	Laughs Shrugs Lights a cigarette	Semi-direct gaze	Medium shot. Personal room, posters in the background	
A	NA	NA	NA	
rontal position, head slightly bent forward (over keyboard)	Smokes a cigarette. Scratches neck	Looks downwards and to left and right	Medium shot. Personal room, posters in the background	
JA	NA	NA	NA	

JA = Data are not available

Appendix 2.2

Speech Transcription

Essence: yeah/ I didn't say nothing/ ah.. improper or .. on his room/ tell me what I break/ because that/ I–I copy the rules/ yeah/ makes the rule/ that's all//

Essence: yes Tod/ Tod you can make/ it SMALL you know/make it bigger/ like on the 16/ 14 something like that/ used for it like this/ y'know//

Essence: well/ you know what? what I like if they use like.. ah.. different/ because if you like chat with someone/ ah.. you know/ you're gonna click an action/ or make see this color or something/ we (wait here)/ I love you by the way/ uhm.. and you just synchrony/ you see this color/ and it's all/ then you know/ that this person likes something in the room//

Sensei: so (. . .) I got this new update/ whatever/ that's cool//

Tinker: good morning Sensei /and goodnight Tea/ sleep well sweetheart//

Essence: hey did you say essence? No/ no/ no/ I know I know/ I wake up like three hours ago/ I don't know. . . and you heard that she woke up/ she this mornin'

King: and you? you hear me?

Essence: well not really/ you have like/ uhm.. maximum 16/ you know/ so in day like/ uhm/ they give you a chance to use like 16/ and you can use it/ uhm .. uhm

Sensei: I just gauged my ear//

Tinker: Sensei /at least you can talk this morning/ you don't have to hang on to the mic//

Sensei: I'm all right/ like it's supposed to be//

Tinker: yeah/ it's all cool/ calm and collected/ this morning/ thank God//

Sensei: it's.. it's my mic really loud/ like cause I–I just downloaded this/ like the new version or whatever/ and it's/ my mic it's saying that's like going all the way around/ like it's really loud or something//

Essence: the (reaction) is good/ if it is called like.. like maximum/ you know/ I dunno (. . .) like ah I can't connect if it's like red/ then and it is like uhm/ it is like all the time/ you know or .. I don't know how to explain it/ but yeah/ IT is on the red and you're hooked//

Tinker: yeah/and Sensei/ if you didn't understand THAT/ Sensei your audio's fine//

Tinker: Sensei/ I'm sticking with five point five/ until Diamond–until he stop usin'it/ then I'll upgrade//

Sensei: uhm .. I see/ uhm.. yes/ pretty cool/ like uhm .. I can show you what it looks like on my cam/ if you wan'it / it's like totally different lookin'//

Essence: it's like a six zero (nine)/ now you have six zero EIGHT/ and by six zero eight it's different/ than .. you can see a cam if there're pro users and/ you know/ if you go on the user/ and nickname like camera/ status/ watching/ age/and ok

Essence: oh then it's nine / yeah/ sorry/ sorry/ it's EIGHT//

Essence: I use Kermit/ I don't know/ I got used to the Kermit//

Essence: I have a feeling somebody's watching//

Tinker: yeah/ go ahead Diamond /what's–what's the problem?

Tinker: no/ I'm not/ I'm single//

Sensei: do you see it? how it's different/ like you were on the (. . .) version is (. . .)/ is on my mic is like going up/ like in a circle/ or whatever/ that's what I was asking if it was too loud or not//

Essence: no/ it's not too loud/ because it's not on the red//

Essence: it's perfect/ yeah//

Tinker: I have no idea/ who is married here/ and who's not married Diamond/ I–I just know/ that I'm not//

Tinker: yeah/ I think engaged people should wear SIGNS/ like toilets do//

Essence: no/ you cannot block that/ somebody cannot view/ right.. uhm

Tinker: have you ever seen/ an engaged sign on the toilet door? I think all engaged people should wear signs saying/ ENGAGED//

Sensei: not married/ thank God//

Tinker: yeah/ ah/ Sensei/ have you tried the new/ ah.. the new ice-ops six one?

Sensei: uhm .. the ice-op .. I don't know what that is//

Tinker: yeah/ it's the operator's tool pack that goes with the/ ah.. Camfrog//

Sensei: oh you mean like Kermit .. or whatever/ that what you're talking about/ I-I never like figured out how to use it/ cause I came in an open room one time/ they told me to download it/ uhm/ and I downloaded it/ I ain't trying usin'it/ I just couldn't figure it out/so ah .. I just did it from/ this .. just Camfrog//

Essence: I don't know/ you've a (choice of many cams) / if you wish//

Tinker: yeah Sensei/ah .. ice-ops are very simple to use/ like Kermit's very simple/ ah.. once you get the hang of it//

Sensei: eh that's ok/ ah/ I don't know /if I'm using it anymore anyway/I'm not in an open room anymore so.. cause there's too much RESPONSI-BILITY/ you know/ I'm just here to chat/ you know/ talk to people//

Essence: (. . .) a great op and fast/ and . . . this not the talk/ you just simply/ ah.. write/ and it's open/ you know/ just click on the right (goof) and it's copied to correct/ his name-nickname/ and you have like a warn/ warn if you wanna warn them/ it is kick/and he kicked up/ and you can make choices like/ what you gonna use like/ ah/ the face on the cam or/ I know you like (. . .)

Essence: I mean/ all you gonna warn though//

Essence: yes/ I did/ Tod/ oh my God/ that's nasty man/ how can you be like that?

Sensei: I can smell it through my speakers (laughs)/ oh yeah/ like I said Tod/
you know/ I really don't care/ if I have that ice-op or Kermit or whatever/
you know/ just keep it simple and basic//

Essence: and didn't he say his volume on or something/ and that'd be inter-
esting/ ah.. for the coarse/ like he's gonna fart//

Sensei: did I say something bad? Because sometimes I–I curse/ without even
knowing that I do it/ because that's just how I talk/ and.. yeah .. sorry if
I did that//

Essence: well /no/you didn't//

3   Writing Out into the Abyss
Polymorphic BlogEng

3.1 BLOGENG

This chapter investigates dimensions of personal/global communication in blogging communities, against the backdrop of globalization and mass culture and under the perspective of language mode variation. *Copy-and-paste* culture radically transforms traditional systems of information exchange on a global scale, and typical features of such a culture may be associated with different language mode dimensions: interactivity vs. individualism, conversation threads evanescence vs. permanence of writing, informality vs. hierarchical control, anonymity vs. authorship, independence vs. mainstream media, credibility vs. opinionated writing.

Prominent in blogs is the constant switching between writing and speech and *vice versa*, all part of the accelerating exchange processes of information management and opinion making, but also part of the ideological foundations of presumed democratic writing and sharing systems of communication. Blogs fluctuate between a personal and intimate sphere (e.g., a private diary) and a public discussion space (e.g., public blogs in online newspapers, cf. Cambria 2011). Does language mode vary accordingly?

The case study presented in this chapter is an exploration into the world of blogging, specifically delving into semiotic integrations and *resource-switching*, which is a distinctive property of digital environments, instantiating alternation of semiotic resources more broadly than mode-switching. Blogs used to be anchored in written discourse, but today their multimodal nature requires specific tools of analysis to tackle the evolution of such texts and communicative strategies within blogging communities. Intersections of written and spoken discourse (e.g., written blog vs. audioblog or videoblog/vlog) emerge along with other semiotic resources, such as the use of visuals (e.g., photoblogs), whereas writing plays a subservient and secondary role (e.g., captions). Analyses of lexical variation in a purposely created corpus of entries from the *LiveJournal* community will also provide insights into language mode variation, assuming that lexical analysis is a powerful tool for the exploration of the spoken/written *continuum*. In this chapter, keyword and lexical bundle analyses are used to probe into a qualitative, quantitative and language-based approach for a comprehensive understanding of the blogging experience that is defined as *polymorphic*, given the instability and rapid changes it is currently undergoing.

BlogEng is thus a label to describe language use in blogging practices and communities, a label that is intended as a far-ranging mould encapsulating language *per se* and other semiotic resources, thus embracing a range of communicative strategies and interactional routines. Multimodal studies are used in this chapter as a full-fledged model to capture the many slippery facets of such web-based textuality that gained popularity as an individual form of communication, but is now in the process of transition toward systems of social networking practices and activities.

3.1.1   "C'mon Join the Blogosphere": Blog Culture and Derivations

A blog is a web page updated on a regular basis with discrete entries (or posts), often delivered to a reader/viewer who has subscribed and who reads it using a secondary interface, or RSS aggregator, rather than directly visiting the blog's website (Rosenberg 2010). An RSS aggregator allows the reader/viewer to combine multiple blog feeds into a single customised web page. A strong link between the producer of a blog (i.e., the blogger) and their audience (i.e., the receiver) is established in blogs (Welch 2003; Stone 2004; B. Miller 2012). Interactivity is a seminal part of the digital world and interactivity has been interpreted as typical of oral discourse in Chapter 1. However, the typical written nature of blog entries further challenges the epistemological nature of such web-based genres.

"Weblogs" or blogs started circulating in the late Nineties in the US, soon becoming an intrinsic living cell of the internet organism (Winer 2003; Walker Rettberg 2008; Rosenberg 2010). In 1999, the fifty or so existing blogs were familiar only to a small group of internet addicts, but by 2006, about 39 percent (57 million) of US internet users regularly accessed blogs, according to the 2006 Pew Internet and American Life Project Report (Lenhart and Fox 2006). They have been broadly defined as web-based journals or diaries, where blog authors, or bloggers, publish their posts in inverted chronological order (Herring et al. 2007). They frequently update their blog, commonly providing comments and criticism about individuals or social themes, thus ranging from the most private (Nardi et al. 2004) to the most public topic, or recording experiences, collecting information for other bloggers, typically supplying relevant links to other websites (Rosenberg 2010). The issue of frequency is of the utmost relevance, as Blood has asserted that "most webloggers make a point of giving their readers something new to read every day" (Blood 2002a: 9). Usually the blogger's aim is *to be read*, so frequent updates generally attract more visitors and help keep a certain amount of daily visitors. Frequency goes hand in hand with rapidity, which is associated with oral discourse features of communication and related questions, such as authorship, storage and control.

According to Herring et al. (2007: 3), blogging is popular partly because it is an easy and inexpensive way to publish materials on the web and reach an enormous audience, and partly because it is more flexible and interactive than previous publishable material, both in digital and print format. They report that blogs have been classified according to external events (known as *filters*, Blood 2002b) and to the impact and influence bloggers may have as *citizen journalists* (Lasica 2002a, 2003; Gillmor 2003), *public intellectuals* (Park 2003) and *opinion leaders* (Delwiche 2004). To put the matter simply, a blogger is an information sharer, a collector of his/her own and others' opinions, always on the move, at least on the web.

The first blog is said also to have been the first website in 1992. However, blogging was not widespread until the late Nineties. Early blogs were mostly lists of recommended links with basic commentary, but soon they evolved into a popular trend, where anyone could write, easily publish, and be read by hundreds or even thousands. The incredible ease of starting up a blog has made it a common and friendly tool in the hands of anyone with a computer and an internet connection. Some bloggers contain personal and mundane observations about "small worlds" or "small events," as Halliday would put it (1978), whereas others are used to make political statements, promote products, provide research information, and give tutorials (Ng 2012).

Since the early 2000s, everyone has been jumping on the blogging bandwagon. Actors, musicians, politicians, painters, novelists, sports people and newscasters, to name but a few, have been launching personal blogs for a variety of purposes, mainly commercial, but also driven by vanity and exhibitionism, as has been lamented in some quarters (Ng 2012).

For this reason, a blog can also be a highly controversial piece of writing. It is rapidly written, published, read and commented on. The well-established and typically long procedures that used to be the norm in the publishing industry have been, if not abandoned, at least frayed. The publishing circle used to reserve the opportunity to publish and be read by an audience made up of intellectuals and acolytes to a chosen elite of other intellectuals and writers. Dropping, at least partially, such habits has had an enormous impact in the medium term, as it will probably have in the long term. Anyone can easily start a blog and write about names and facts that may become public in a matter of seconds. In effect, from this standpoint, blogs can become double-edged weapons that can be turned against their own creators, as in the case of employees being fired after their criticism of or embarrassing reports about their employers on their blogs. Some bloggers are anonymous or use a pseudonym for privacy. Others share their blogs with friends, family and co-workers, even though information, if negative or damaging for one's own image, can be potentially harmful. Questions of provenance thus become crucial, and authorship can be claimed or otherwise camouflaged or rejected to these ends (Qian and Scott 2007).

At the beginning of the twenty-first century, blogs have become so widespread that the word *blog* was Merriam-Webster's word of the year in 2004. Families use blogs to exchange daily information, and university and school teachers use blogs as writing assignments. *Newsweek* magazine even recommends a few blogs each week to its readers (Ng 2012). A number of blogs may also feature advertisements to earn some extra income or free products, while some businesses use blogs on their websites to establish a more personal relationship, based on interaction with their prospective customers. Some bloggers use personal sites to post music or photos, as in the case of photoblogs, which often do not include text in the entries, with the important exception of captions that are in many ways subservient to the primacy of images. These are real visual statements, to borrow from Kress and van Leeuwen (1996, 2006).

Despite the general interest on the part of bloggers in *being actually read* by as many people as possible, it would be misleading to consider blogs and bloggers as a uniform and homogenous category. Bruns and Jacobs (2006), for example, reject the narrow view of blogging as writing reverse-chronological posting of individual entries, with the inclusion of indexed hyperlinks and response entries. They contend that considering blogging as if it were a static and all-embracing category was likely to be abandoned in research literature, also reflecting the need to explore specific blog genres, such as diary blogging, corporate blogging, community blogging and research blogging (Bruns and Jacobs 2006).

3.1.2   State of the Blogosphere

The most recent Pew Report on social media and mobile internet use (Lenhart et al. 2010) shows that blogging has decreased among teens and young adults since 2006. However, among adult users (age range twenty-five to thirty-four, according to *Technorati*, 2011), interest is increasing. Specialized search engines or meta-directories such as *NM Incite* or *Technorati* have tracked over 181 million blogs as of December 2011, up by thirty-six million from 2006 (NM Incite Report 2012b). Three of the top ten social networks are blogging platforms, such as *Blogger*, *WordPress* and *Tumblr*, with a combined eighty million unique visitors (NM Incite Report 2012b), even though the exact number of blogs is impossible to determine, because a considerable number are abandoned after a few weeks or days. Furthermore, the nature itself of the blogosphere is networked and decentralized, thus making precise estimates impracticable. According to mnincite.com, more than six million users write blogs using blogging services and an additional twelve million users publish blogs using social networking websites (NM Incite Report 2012a).

As social networking practices have become more interactive and the technologies and tools embedded are constantly changing, teens and young adults may prefer *micro-blogging* instead of traditional blogging.

Micro-blogging refers to systems of updating status in social networks, such as *Facebook*, *Twitter* or *Reddit*, and it is quicker and more efficient in disseminating a writer's ideas, opinions and quotes, and as such, has earned huge popularity in recent times. Lenhart et al. report that in 2010, 14 percent of teens claimed they blog, down from 28 percent in 2006 (2010). Teens now blog only marginally more than adults, about 12 percent of whom have reported blogging since February 2007. The significant drop in teen blogging is also consistent with the reduction of teens commenting other people's blogs (or within social networking platforms) that in 2010 was 56 percent, down from a striking 76 percent in 2006.

As in recent years, teens from lower-income families are still more likely to keep a blog, and no significant statistical difference in male/female blogging is currently reported, as well as no noticeable ethnic differentiation among blogging teens. However, according to the Pew Report (Lenhart et al. 2010), one in ten online adults persist in keeping a personal journal or diary, thus increasing the amount of blogging among adults, from 37 percent in 2008 up to 47 percent toward the end of 2010. Reasons why blogging has dropped may be explained in terms of the use of social networking that has steadily increased in the past few years. Nearly three-quarters of teens (73%), and an equal number of young adults (72%), use social networking platforms, whereas only 40 percent of older adults (age thirty and older) use them.

As Lee Rainie, director of the Internet & American Life Project, has claimed, blogging is not so much dying, but shifting with the times: "The act of telling your story and sharing part of your life with somebody is alive and well—even more so than at the dawn of blogging [. . .] It's just morphing onto other platforms" (Kopytoff 2011). The example of *Tumblr* is particularly relevant: although this platform is embedded in a blogging service technology, it shows that its users may be unaware of the differences between blogging and sharing other semiotic resources, texts or web genres, for example posting photos or videos.

This raises the question as to whether the decline of blogging is merely a *semantic shift* to other forms of interaction and ensuing deployment of other semiotic resources, where writing is just one among many. This semantic shift, however, impinges on major blogging services companies, such as *Blogger* or *LiveJournal*. The latter has placed more emphasis on social and interactive communities in recent times, which can be interpreted as a shift to social networking practices. Sue Rosenstock, a spokeswoman for *LiveJournal*, significantly commented on the matter: "Blogging can be a very lonely occupation; *you write out into the abyss*" (Kopytoff 2011, emphasis mine). The "abysmal" quality of writing resurfaces here and is additionally characterized by other features, such as the time and effort necessary to write full-length blog entries. Full sustained prose found in traditional or *macro-blogging* has been abandoned by younger generations, as the 2010 Pew Report suggests (Lenhart et al. 2010), as they now appear to be favouring micro-blogging alternatives, such as those constituted by *Twitter* or other status update systems.

Blogging, however, is today more associated with personal and opinionated writing style. But *how* is this writing style different and how are these differences deployed? Much research on mass communication has focused on the sociodemographics of bloggers and individuals' motivations for using a specific medium (Papacharissi 2004; Herring et al. 2005a, 2005b; Kaye 2005; Nowson and Oberlander 2006; Li 2007; Sanderson 2008).

In particular, what are the motivations behind blog writing? What are their perceived and real benefits? Studies exploring such issues deal with uses and gratifications deriving from media use and have found that users generally employ internet for information-seeking or entertainment. Examining blogs and establishing connections between its uses and gratifications, Kaye maintains that blog usage was a predictor, but internet experience was not (Kaye 2005). This suggests that the effects of blog usage may not be parallel to those of other media, in particular with regard to credibility, as media use has also been found to be a predictor of credibility of *that* medium. Research literature has also found that perceived credibility in the media has been influenced by gender, type of medium, expertise in the medium and motivations (Armstrong and McAdams 2009). Credibility is a fundamental predictor for understanding why blogs have steadily challenged mainstream media, as users tend to abandon media that are not perceived as reliable and producing thoughtful analysis (Kaye and Johnson 2004a, 2004b; Kaye 2005; Johnson et al. 2008). Another reason has been found in the commonly shared view that they are independent of corporate-controlled media.

Credibility is, of course, linked to notions of authorship that, as discussed in Chapter 1, is peculiar to web-based communication as the author, or the person recognized as the author, is not as straightforward as it used to be in spontaneous, live interaction or in mainstream, traditional media. The blog author is generally held to be the "source" of information and perceived credibility is attached to this assumption of recognizable authorship, along with considerations connected to provenance and storage. To this end, blogs, especially filter blogs, usually link to other websites to corroborate the authenticity/credibility of their claims. This system may be associated with the long-standing habit of quoting sources in traditional written genres. In blogs, as Armstrong and McAdams note (2009: 441), those links are often to mainstream media or information-oriented websites; however, they are not systematically checked for credibility. Notions of authorship are important in as much as they deploy the personality of the writer, which is deemed a fundamental quality of blog writing (Winer 2003). This notion is also connected to traditional views on written communication, where author and provenance are clearly identifiable. However, far from being bound to source reliability, blog credibility has been found to depend also on *purposes* for blog reading, with information-seeking blog readers more intent on lending credibility. Most importantly, this study has found that the subject itself and the writing style was essential in lending credibility, with the interesting finding that a more cynical and opinionated style may be perceived as more reliable and trustworthy (Armstrong and McAdams 2009). However, studies of mainstream media suggest that opinionated writing lowers medium credibility (Metzger et al. 2003). Blog readers value the opinionated style of blog writing as more trustworthy than the presumed "objective" style of mainstream media (Lasica 2002b; Bruns 2006).

Connected to these issues is the relationship between blogs and mainstream journalism (Lasica 2002a, 2003; Blood 2003; Haas 2005; Tremayne 2006; Carlson 2007; Kenix 2009), blogging and politics (Park 2003; Kaye and Johnson 2004a, 2004b; Adamic and Glanch 2005; Auty 2005; Bahnisch 2006), corporate blogging (Charman 2006; Kelleher and B. Miller 2006; Scoble and Shel 2006), blogging interpersonal features (C. Miller and Shepherd 2004; Nardi et al. 2004; Stefanone and Jang 2008; Thibault 2012), and blogging and gender (Herring et al. 2004a; Herring and Paolillo 2006; Pedersen and Macafee 2007).

Placing the practice of blogging in a cultural dimension and considering that blogs are mainly constituted by language use, an extensive body of research into blogging has identified English as the blog's *lingua franca* and this book is no exception. However, other studies have investigated other cultural contexts, bringing together this and other strands (Rogers and Marres 2002; Merelo et al. 2004; Tateo 2005; Esmaili et al. 2006; Hodkinson 2006; Miura and Yamashita 2007; Xiao 2009).

From this general outline, we may well argue that blogs cannot be easily defined. As Schmidt claims (2007), we can speak of blogs only in very broad terms. They are so interconnected to other texts and genres that it appears almost inconsistent to define a "blog" as an autonomous genre, especially if we consider recent changes that multiply blog genres and blogging practices and culture.

3.1.3   Amateur and Professional Writing: Blog and Mainstream Media

What kind of relationship is established between filter/politics blogs and mainstream media? It has been argued that filter/politics blogs are an extension of 1960s *New Journalism* and, as such, they represent a form of alternative media (Kenix 2009). But are these blogs really an alternative to mainstream journalism? There are no straightforward answers to this question. If we tackle this matter in relation to this book's agenda, connecting spoken and written discourse to digital environments, the relationship between media and the preferential language mode/s is a useful heuristic model for study. However, binary notions are not useful for exploring the relationship between blog and mainstream journalism.

If we accept that blogs are abstract versions of independent and alternative journalism, they can also be seen as uncensored and unrestrained reports on news, drawing on a range of different sources, and encouraging a free and egalitarian exchange of information. This view clearly puts blogs at an advantage when compared to corporate, hierarchical and mainstream forms of journalism, which is conditioned by profit requirements and company agendas. As multinational and profit-centred corporate giants own and direct mainstream outlets, the proliferation of blogs may be explained in terms of the need for so-called *citizen journalism*, which has been nonetheless challenged and questioned in a number of studies on the matter (Gillmor 2003; Haas 2005; Kenix 2009), also in the debate on the relationship between the two idealized alter egos, that is, the amateur blogger and the professional journalist (Rosen 2011).

Despite their attempts at being "more credible" than mainstream journalism, blogs have been found full of highly controversial and opinionated claims, even though some practices, such as hyperlinking multiple sources, counterbalance such implied biases: "[U]nlike static websites, blogs depend upon hyperlinks not only to boost attention to their own blog, but also to ensure that users can be quickly led to relevant information," also providing time and space to think critically and offering "an antidote to the mass-mediated, corporatized culture that surrounds us" (Kenix 2009: 792).

Quoting other blogs and including them in *blogrolls* is a typical activity that adds to bloggers' struggle for credibility and also creates an *echo-chamber* of multiple cross-referencing that ultimately results in the *A-list* for blogs (Adamic and Adar 2001; Kenix 2009). However, research has found that bloggers primarily include *like-minded* blogs in their blogroll. Interestingly, when the A-list is at stake, bloggers prefer to link to mainstream media than to non-corporate outlets (Adamic and Glance 2005).

Although research literature on the matter indicates that blogs are less genuine and alternative with respect to mainstream media than was commonly supposed, they nonetheless play a significant part in providing accessible outlets for influencing, if not steering, mainstream media. They make available free and uncensored reports on news and also on bloggers' private lives in the case of personal blogs. Just as with online newspapers (Knox 2007, 2009a, 2009b; Cambria 2011), blogs can become corporate and institutionalized and this happens frequently with the most visited and rated blogs (see the case of *The Huffington Post*, according to *Technorati* 2011 the most visited at the moment of writing this chapter). As a sign of the change in mediascape, two online news outlets, *The Huffington Post* and *Politico*, won their first Pulitzer Prizes in 2012 (Chozick 2012). Furthermore, corporate online media often embed blogging practices in their own websites, as is the case with most English or American language online media.

Blogging can thus represent a chance of democratic and participatory sharing of contents (Jenkins 2006), but a more accurate view into such phenomena may shed light on multimodal interactional practices in digital environments. Herring et al. (2004b: 6), who explored the "asymmetrical communication rights" between bloggers and commentators, observe that bloggers "retain ultimate control over the blog's content." In effect, they highlight the power bloggers have to *moderate* comments, thus ultimately drawing in the reins of what is visible or not. As a consequence, we may postulate that such a direct control over publishable materials puts bloggers in the position of undemocratic and authoritarian leaders at the end of the day. It is nonetheless true that the possibility of easily becoming a blogger or posting comments on a variety of topics in diverse outlets is unprecedented in the history of mainstream media. The albeit virtual possibility of posting comments is doubtless a way of encouraging and making available practical systems of participation and of content sharing. Hence, this possibility parallels the vicarious creation of a sense of community in digital environments.

3.1.4 Blogs and Their Position in the Digital World

As stated in the previous chapter, communication among parties in video-chat is established, on a basic level, within the inner circle of family, relatives and friends, or extended to the outer circle of work environments. This inner circle is made up of closely related people or acquaintances. This kind of video communication is set up via web-based software programmes and is most typically made up of one-to-one (or, occasionally, one-to-many) video verbal exchanges or full video conversations. The length, number of participants and language used is highly variable, but a typical feature of this kind of video interaction is the fact that participants are either likely to know each other in advance or are likely to meet outside the digital world. Of course, we could mention exceptions, but, typically, one-to-one conversations are characterized by mutual acquaintance in the web-based environment. A different kind of video interaction is represented by multiparty video interactions, as in the case of Camfrog or similar multi-room video-chat software programmes. In this type of web-based interaction, participants are not likely to have met in advance, mainly because the maintenance of previous social bonds (such as those with family, friends or co-workers) usually does not imply the use of such open environments, where any external observer may join in or intrude. Such video communities are based on the idea of meeting and making new friends, as the thematic "chat rooms" seem to suggest. For example, chat rooms such as *SpeakEnglish*, *HableEspanol*, *Single* or *Sign Language* in Camfrog imply a specific audience, made up of a specific community of participants and, more broadly, of practice, for example those speaking a certain language or having a common demographic independent variable (e.g., being single).

This is a seminal difference that comes into play in the definition of the second web-based environment discussed in this chapter, that is the (we) blog. Blogs cannot be defined by a standard, preferential or definitional use of a specific semiotic resource. As we have discussed in the previous chapter, videochats are ingrained in a web-based environment, where, as the word itself suggests, the video is a definitional feature. A videochat may be interspersed with, even mostly constituted by, written comments, but the presence of the video is, by definition, a default feature of such a web genre.

Conversely, blogs are not defined by a single semiotic resource. As we will see in further detail, blogs are subdivided in other genres, such as the *videoblog*, *photoblog* or *bleg*. However, any blog usually blends more than one resource. Historically, and relatively speaking, blogs are made up of written words. As research literature attests, blogs have been compared to mainstream journalism and many issues have been tackled with reference to logocentric questions, such as issues of credibility, relationships with mainstream media, language use with reference to dependent or independent variables (e.g., bloggers' demographics) and, more generally, blogging practices and systems of communication. All these questions, however, are centred around the notion of *planned written communication*, with scarce reference to other blogging systems, for example with reference to unplanned (either spoken or written) communication or communication exclusively based on visual language. However, investigation in this respect is needed to fill these gaps in research literature.

Another fundamental difference deals with the reciprocal status of participants. Whereas in web-based video interaction, participants have the same status (i.e., they can in theory take turns as in real-life conversations), in blogs, participants have a different status. In particular, the blogger, who is the top hierarchical source irradiating contents, displays, by definition, a higher status compared to the audience. Even assuming a voluntary *low* degree of control on the part of the blogger, for example by granting commentators on the blog unlimited access to contents and unmonitored commenting activity, such a hierarchy is nonetheless there, as it is the blogger who has the power to define and set the blog's rules. The blogger has the power to determine which contents are available, and to whom and to what extent they are arguable, so that the definitional unequal status relationship between bloggers and their audience is established.

Blogs are complex to define, as this discussion has briefly documented. They cannot be studied as if they were a unique genre. On the contrary, there are so many different kinds of blogs that it is almost impossible to find ways of categorizing them. The early definition by Herring et al. (2004), which postulated the existence of at least three sub-genres, such as diary blogs, filter blogs and k(knowledge)-blogs, is still valid today. However, it is also very likely that a blog is made up of several components, thus making it virtually impossible to fit it into a unique/univocal category.

3.1.5   "You Can't Just Simply Say: It's Bloggin'!": Polymorphic Blogging

As argued in the previous section, blogs take on different forms and shapes. In this respect, they are one of the most polymorphic genres and change so frequently that even the very notion of blogosphere may become almost meaningless in a short time.

Even a cursory look into social platforms providing blogging services, such as *LiveJournal*, affords such a complex mosaic that it justifies the fear of abyss suggested in this chapter's title. This fear has something to do with the fear of chaos, where every definition loses its ground and no common rule of description can be established. As said, blogging is being transformed by the contribution of several semiotic resources, each adding to a portion of meaning that needs to be unpacked and interpreted in their mutual orchestration.

The videoblog (henceforth, vlog) is an emerging genre and, as suggested in section 3.1.4, it can, in some ways, be compared to the videochat, especially if we consider the semiotic presence of the video. But what is a vlog, and who are vloggers?

The term *vlogging* combines *video* and *blogging* and places the video as the foregrounded semiotic resource. Accordingly, a vlogger is a vlog's author, who posts videos as entries to their blog instead of a written text or other leaner semiotic resources. Videos can also be accompanied by verbal text or other semiotic resources, such as photos or drawings. As a common practice, vloggers post videos that feature themselves, and they frequently play some major part in the video's production process. Vloggers may also post videos related to a specific genre, such as fanfiction, sports, or other hobbies or interests. The distribution of vlog updates is often facilitated by an RSS (i.e., Really Simple Syndication) feed, sometimes also referred to as a *vidcast* or *vodcast* (Anissimov 2012; Melanson 2012).

The first vlog is generally considered to be "The Journey," by Adam Kontras in 2000. It chronicled Kontras's move to Los Angeles and subsequent attempt to break into show business. Kontras's blog was applauded as a significant breakthrough for his use of the medium. Since then, the vlogosphere has considerably expanded and vloggers now have their own annual conference, *Vloggercon*, and their own vlogger awards, *The Vloggies* (Anissimov 2012; Melanson 2012). Although the early 2000s were marked by attempts to create videoblogs, vlogging did not truly take off until 2004, when small communities of vloggers began to emerge and mainstream media started to notice this new digital experience. Vlogging is not yet widespread, but online video is certainly becoming hugely popular, with websites such as *Google Video* and *YouTube* offering free storage space (Anissimov 2012).

A considerable number of vloggers post videos directly to their vlogs via *YouTube*, which is the most popular video sharing community, with an audience of over seventy million unique users per month. Vlogs are usually created using Web 2.0 tools and services such as *Joomla*, *LiveJournal*, *WordPress* and *Blogger*, which allow the posting and embedding of videos via content management systems (CMS). The development of digital recording affordances of mobile devices has also contributed to the expansion of vlogging practices. It is believed in some quarters that vlogs will soon replace static text and images, also due to the adoption and falling cost of broadband technology. Bloggers can also use different semiotic resources, thanks to RSS enclosures, allowing distribution to the aggregators of blog's subscribers (Anissimov 2012). Despite the presence of oral discourse in vlogs, they can hardly be identified as web-based forms of spontaneous conversation, as videochats are. Vlogs are reproductions of what bloggers could have written, as they are mostly examples of *planned-in-advance written texts*, delivered by speech and reproduced through digital recording. From this standpoint, they are similar to their written counterparts, and changes in resource do not significantly alter their intrinsic properties.

Another sub-genre is the *bleg*, a portmanteau of *blog* and *beg*, that refers to a request for assistance posted by bloggers. In some cases, an entire website can be a bleg, maintained specifically for the purpose of asking for something, and in other instances, a blogger may publish a blegging post to ask for help in special circumstances. People may post blegs for a number of reasons. For example, some bloggers ask for help when they need to buy new blogging-related equipment, or to raise funds on behalf of an organization or individual. Sometimes, a bleg does not involve financial or material support. For example, a blogger might bleg for information about something, like updates from people in a disaster area, or commentary from readers about topics of interest (Smith 2012), as will be detailed in section 3.2.3.

3.2 A METHODOLOGICAL OVERVIEW

How are speech and writing used in blogs? The previous sections would suggest that speech is used in vlogs and writing in traditional blogs. However, as Thibault (2012) maintains, the possibility of hypertexting in digital media has reshaped language, and in particular *written language*. Rethinking textuality is a major focus in this book and Thibault (2012: 15) discusses this topic in the following terms:

> Perhaps the biggest casualty of this re-thinking is the notion of "text." In the Western tradition, text has been seen as a determinate object or output of a particular linguistic production process. Both literary theorists and linguists have reified "text" as (1) a stable linguistic artefact that can be consulted and inspected by multiple users on different occasions and (2) which is deictically anchored to some kind of authorial responsibility and sponsorship (Harris 2000: 236). This has been equally true of both speaking and writing. Linguists and discourse analysts have used written transcriptions of spoken events as a kind of guarantor of stability of the object of study, viz. a dynamical, time-bound event. As with written documents and records, the writing system was conflated with the notion of text such that the former was taken to be the guarantor of the stability and invariability of the latter. On this basis, texts

were taken to be coherent and stable entities on analogy with the material organization of the written script that supports them.

From a theoretical standpoint, polymorphic blogging is a definition that exempts us from an excessive trust on traditional notions of genres. Polymorphism is defined by the Merriam-Webster Dictionary as the "quality or state of existing in or assuming different forms." This property applies to blogs, as they are undergoing the fastest change in their evolution so far. It is far from straightforward to find a single definition for blogs and identify them with one particular web genre. The notion of polymorphic blogging assumes genre evanescence, similar to that which we discussed in Chapter 1 about stereotypical forms of speech. However, the attached notion of *planned-in-advance writing*, that is fundamental in most blogs, obscures the question.

From the previous sections, we have come to the conclusion that among the main concerns of research literature, we find issues such as the initial identification of blogging practices with so-called *citizen journalism* and relationships with mainstream media, with ensuing questions, such as reliability and credibility of each medium. Furthermore, blogs call into question notions such as authorship and control, linked to the chosen style of writing, from the personal, informal and casual to the public and reliable.

Following some considerations from the previous chapter, we may well argue that identification of this genre is controversial. Thibault (2012: 12) rejects the view of genre as a mere "template" to achieve a specific communicative goal, as, he argues, it is a "kind of discursive 'glue' that holds other things together in a coherent cultural format."

Labelling blogging as a polymorphic genre thus tries to capture this complexity and rapidity of change. It also reflects several different practices that have been so far treated as uniform and homogeneous in some research literature. For example, *macro-* and *micro-blogging* are two labels that are spreading in digital platforms, such as *Twitter*. Micro-blogging has come to the forefront of blogging practices, according to several recent reports. Macro-blogging is the "oldest" practice in relative terms, implying the use of fully sustained prose in a closed or semi-closed digital environment, whereas micro-blogging is the most recent and less-explored practice, involving the use of concise entries that are becoming increasingly embedded in wider interconnected blogging platforms. They are disseminated through systems, such as blogrolls, and promoted by profit-based and corporate ideologies, such as the *A-lists*, nurtured by popularity and control.

In this chapter, we will review *macro-blogging as it is merging with micro-blogging*. Our analysis will thus need to incorporate all these strands.

The methodological steps of this chapter include:

1. The use of purposely created corpora, with different groups of datasets and access to specific communicative exchanges in blog entries.

2. The inclusion of several semiotic resources in such corpora, allowing quantitative and qualitative analyses that compare verbal and non-verbal language use.
3. The use of specific software programmes (i.e., Wordsmith 5 and 6) to count frequencies and patterns of co-occurrence of linguistic features that have been selected as symptomatic of the written-spoken *continuum*, mostly on corpus-based lexical observations.
4. The application of a corpus linguistics approach to study multivariate dimensions of language use and variation in the sample, with particular reference to both positive and negative keyness in comparison to some reference corpora.
5. Discussion of the most frequent lexical bundles, which are considered indicators of register variation (Biber and Conrad 1999; Biber 2003; Cortés 2006).
6. Separate but combinable analyses for different semiotic resources, such as verbal language, images, videos and others, in order to take into full account how multimodal resources come into play in digital communication, namely in blogging communicative events and how resources combine to create meaning-making events.

3.2.1   Step 1: Corpus Design and Text Selection

The initial step involves the choice of texts to be included in the sample, also considering issues of representativeness and the selection of linguistic and non-verbal features to be included for analysis (McEnery and Wilson 2001). The question of representativeness of the blog corpus has been computed, on a lexical level, by the degree of "saturation" (Belica 1996) for specific linguistic features (e.g., lexicon size). Saturation of blog language means that the linguistic features show very little variation. To measure corpus variation, the corpus has been divided into several segments of equal size, based on its tokens. The corpus, in the form used for the present analysis, can be considered as *saturated* on a lexical level, because each addition yielded approximately the same number of new lexical items as the previous segment (cf. section 3.3). Saturation for representativeness has here been preferred to notions of balance, but it must be noted that saturation is significant only on a lexical level (McEnery, Xiao, and Tono 2006). However, corpus balance has also been taken into account. As the corpus is specialized and, as such, includes only blog entries, there was no need to balance features such as text types or genres. Entries have been collected randomly, provided their length was greater than or equal to one hundred words.

The present study is based on a *specialized dynamic monitor corpus* (cf. McEnery, Xiao, and Tono 2006), in much the same way as was done with the video corpus presented in the previous chapter. Both corpora may be increased in size and also become more balanced, for example for the purpose of undertaking a longitudinal analysis that is, at the moment, beyond the scope of this book. In the case of the blog corpus, a "time-frozen" analysis has been preferred, to capture *a specific moment in time* that is proximal to the moment of writing this book.

Closely related to notions of representativeness and balance is the underlying procedure of sampling, whereas *sampling units* are represented by *blog entries*. In corpus design, special care was initially attempted to take note of the bloggers' demographic variables, such as gender, age, language(s) spoken and education, where possible. However, such data were unfortunately often missing or impossible to verify, but data sample has not been modified accordingly, for the specific purpose of not altering a typical digital environment, where personal details are often camouflaged or missing.

Defining demographic variables and constructing a sampling frame is particularly difficult, and especially so for spoken language, for which there are no readymade *sampling frames*, such as indexes, catalogues or directories (McEnery, Xiao, and Tono 2006). Consequently, selection has been carried out through simple random sampling, also considering it to be, by corpus definition and initial design, quite close to a stratified random sampling, based on homogenous groups or, in this case, texts (i.e., *strata*).

A second consideration in corpus design was the text and related discourse genre collected, which, as said, was restricted to blog entries in a single digital platform, namely *LiveJournal*. Such restricted data samples have been selected with the purpose of observing purely linguistic and multimodal features in a specific blogging community, assuming that this community is sufficiently wide to allow representative linguistic behaviour and sufficiently self-organized as to guarantee some basic technical (e.g., templates) and semiotic (e.g., use of multimodal resources) characteristics for observation.

With regard to corpus size, some preliminary issues have been addressed (cf. McEnery, Xiao, and Tono 2006). How large did the corpus need to be in order to be considered representative? Did full entries and comments need to be included as dataset, or not? As the book's main interest was investigating the role of spoken and written language in interaction with other nonverbal resources, full entries have been included, featuring a pre-established number of comments (≥10). Considering that the dataset was constituted by a sample of entries collected in a very restricted time range (April 2012) according to the criteria of popularity, and that the verbal and multimodal features were, broadly speaking, from frequent to very frequent, the sample was deemed sufficient enough to explore such features.

Biber (1993) maintains that frequent linguistic features are rather stable in their distribution and, consequently, that short texts (about 2,000 words) are usually sufficient for the study of frequent linguistic features, whereas rare features tend to vary to a greater extent in their distribution and, as such, require larger samples. Following this observation, the present corpus presents a small slice of the blogging cake, hopefully significant in its compactness and relative variety. In conclusion, even though a demographic sampling was attempted, a context-governed sampling was ultimately preferred, establishing the criterion of popularity as the governing principle, as this has been singled out as one of the most significant principles for the blog genre and also within one of the most varied blogging communities. The choice to restrict sampling to one platform, despite its doubtless drawbacks, was made to restrict some dependent and independent variables, especially those related to language use and affordances constrained by the medium.

In keeping with the book's aims, separate multimodal corpora, loosely intended as collections of multimodal texts, were built and tagged, exploring different aspects, both strictly linguistic and others more comprehensively semiotic in a broader sense (e.g., text, video, images). One thousand five hundred entries were collected in the selected time span, excluding entries with less than one hundred words and without comments. The selected sample includes all the resources employed by bloggers, including written language, pictures, drawing, animated gifs, videos and others.

3.2.2   Step 2: Corpus Annotation

With regard to annotation, as discussed in the previous chapter, a very specific kind of annotation was undertaken. In Chapter 2, we noticed that several hundreds of hours of recording interaction can be annotated more efficiently and reliably if highly specific research questions are addressed and if corpus annotation, especially if videos are concerned, addresses such questions in a direct, unambiguous and replicable way.

For these reasons, the most efficient way of annotating the corpora in this book was the *problem-oriented* approach. A problem-oriented corpus does not cover all linguistic (and semiotic) phenomena, and is not intended as an exhaustive way of addressing the texts included in the corpus. It is oriented toward one or more specific research questions. Annotation is, thus, restricted to relevant phenomena and can be more easily handled, especially in cases where a full annotation would be extremely time-consuming and virtually impossible to carry out manually. The corpora used for this book are thus highly situated and do not assume attempts at objectivity, as I believe that any kind of experiment—be it in soft or hard sciences—is always situated and inevitably biased.

I have sampled some specific phenomena and discussed them from my highly situated viewpoint, in much the same way as what happens in digital ethnography, but with the support of a larger amount of data and datasets. Furthermore, this kind of annotation fills some gaps that may be found in existing corpora: for example, the *Birmingham Blog Corpus*, which, to my knowledge, is the largest blog corpus, tagged consistently and completely, but which fails to address those features that are relevant to my work. However, this system of annotation is very specific and, as such, can only be used to explore the research questions raised in this study.

In the previous chapter, the main issue was the purely linguistic notion of mode-switching, which was explored through blending corpus linguistics methods, multimodal annotation and conversation analysis. However, the notion of mode-switching can be expanded and include other kinds of switching, not exclusively dealing with the alternation from oral to written language and *vice versa*, but also with *alternations from one semiotic resource to another*.

Mode-switching is, in my theoretical framework, closely connected to Halliday's original identification of mode as the channel of communication, whether spoken or written. As such, mode-switching is a linguistic phenomenon restricted to some specific digital genres, namely the videochat, video conference or *in praesentia* video communicative events in general. As discussed in the previous chapter, this phenomenon is not only circumscribed to this specific digital genre, but it is also quite rare, and nonetheless significant because it marks an unprecedented alternation of speech and writing at the same time and in the same communicative event (Sindoni 2012). It also enacts interesting and rather unexplored patterns of alternation. My effort has been to give a theoretical framework for the enterprise of studying such alternations (i.e., mode-switching) in digital communicative events.

However, in other digital environments, such as blogging platforms or video sharing communities, other mode-switching phenomena may be observed, and will be discussed subsequently. For the moment, we may postulate an even *more* restricted use of mode-switching *stricto sensu*, namely only in the case of vlogs that associate a video output to a co-textual use of writing. By "co-textual use of writing," I mean the visible display of some pieces of writing, for example if vloggers show themselves while typing *visible* written words on their computer, or sending visible written texts to their audience while they speak. The practice of associating videos, and hence speech, with written comments is in effect taken for granted in digital environments and cannot be defined as mode-switching, as this notion assumes the use of speech and writing at the same time (in synchronous mode) and during the same communicative event. It is patent that synchronicity parallels the uniqueness of the communicative event in which it takes place.

To put the matter in simple terms, we have always used speech and writing simultaneously, for example writing a shopping list while speaking on the phone. What is new is the contextual (same time) and co-textual (same communicative event) occurrence of speech and writing. Accordingly, it is pointless to define mode-switching as the common simultaneous use of speech and writing, as is common practice within and also outside the digital world. Mode-switching thus requires the alternation of speech and writing in the special conditions described. Switching from one resource to another will be accordingly considered and computed as a *separate* kind of occurrence, here called *resource-switching*. The latter is different from mode-switching, in ways that have been described here and in the previous chapter.

Following these initial guidelines, two separate corpora have been created. The blog corpus includes only verbal elements (LJC), whereas blog corpus 2 includes also other semiotic resources, such as images, pictures, drawings, videos, animated gifs, and so on (LJC_SR). Blog corpus 2 has been annotated following the problem-oriented approach, hence annotating the entry's title, the entry's subtitle (if present), the entry's full text, and the presence and number of other semiotic resources, including positioning and relative number compared to word number and feeding demographic information, where present, about the entry's author.

These comments serve as a proviso to the following discussion, as blogs and social network practices in general feature mode-switching in a rather restricted sense and also with relative rare frequency ($\leq 10\%$). So how can the interplay of speech and writing be monitored, and which texts and linguistic features are preferable for such monitoring?

A methodological decision about the linguistic features to be selected for analysis was made by revising previous literature on the matter, also outlined in Chapter 1.

3.2.3   Blog Categorization

With regard to the kind of blogs that have been observed in this study, they have been loosely categorized after the observation of more than 3,200 blogs, as detailed in Table 3.1, with specific reference to the *LiveJournal* (LJ henceforth) blogging community.

However crude this categorization may appear, it outlines the major trends in the LJ blogging community. Most blogs actually blur into social networking communities and the distinction between this community and

Table 3.1   General categories for blog types, main function and related examples

Blog type	General description	Example
Fandom and gossip (music, cinema, TV shows, cartoon, digital games)	Fans tell stories about their music, movie, TV or digital hero icon	Oh no they didn't Arama they didn't Oh no they didn't – Glee Oh no they didn't – Star Trek Anything diz Top Model Milady Milord
Fanfiction	Fans write stories based on original books/TV shows/etc.	Fandomsecrets Once upon a meme Kizuna exchange Supernatural Gen Fic Exchange Write & Gripe The Original Psalm

Sport and leisure (e.g., football, cooking, riding, travelling)	Users tell stories about their favourite hobbies, sports events and leisure time, such as cooking, travelling, favourite activities, cities or countries, etc.	Cooking bakebakebake Bad riding Melbourne Maniac Damn Portlanders
Tutorials, icons	Users publish tutorials or self-produced icons	Rapture Icons Come delight Stay lost
Self-help (parenting, work)	Users of a specific community ask and answer questions on community-related topics	Childfree Parenting 101 Customers suck Co-workers suck
Personal life	Users share stories about their lives, often documenting them with pictures	Aamalie archive
Politics	Comments on political issues	Oh no they didn't – Political Politicartoons
Pictures, images	Users share photos or pictures without commenting. Words are only used for captions	Abandoned places Picturing food Lj_photophile
Collectors	Users look for and sell collectables	Pkmncollectors
Ask for general information (bleg)	Users ask general questions, ranging from universal (e.g. love, life) to very specific (e.g. a book's title, a translation)	Little details Linguaphiles Ask me anything The Question Club What was that book

others is not always clear-cut and several digital genres are no longer easily distinguishable, as they seem to merge along a *continuum*. LJ promotional slogan advertises it as a "global social media platform," boasting 56 million journals to date and 138 thousand posts "in the last 24 hours."

Such communities, however, largely rest on blogging practice foundations. Blogs include entries published in inverted chronological order and comments are placed below each entry. Users aggregate mainly around common interests, school or ratings, as the search menu suggests. From this standpoint and following these changes, blogs are not isolated websites featuring personal homepages, but conversely coalesce around notions such as sharing, meshing up and socializing contents, much in the same way as happens with social networking communities, as will be discussed in the following chapter.

Common interests are at the core of blogging communities. Diary blogs or journals also need to appeal to a large audience to be read, commented, rated and blogrolled. Popularity is a central notion for this text genre, as survival, in blogging terms, rests on being read. Broadly speaking, interests aggregate people within and outside the web. However, becoming familiar may be far from easy for an outsider: the apparently open and democratic nature of such communities is controversial. In my personal experience, being rejected several times and by several communities was more than common, when I was trying to get first-hand information on blog practices, on the grounds that I did not belong to the community, and, as such, not entitled to see into its mysteries. Communities can, nonetheless, be a very welcoming space for people sharing some common ground in their non-digital life. Participation is, in fact, an active way of enriching blogging experiences, as is made clear by the desire to be commented expressed by bloggers, when they are not involved in corporate blogging as well.

Before a quantitative analysis of LJC, a qualitative discourse analysis will be introduced to help address the research questions discussed in this book. For this reason, the devised categories—however tentative and incomplete—will be further explored in an attempt to give a general outline of some of the observed trends in recent blogging practices in the selected community, namely LJ. Note that the following only represent qualitative comments, based purely on recurring discourse patterns.

The category listed under the heading of "Fandom and gossip" may be represented by one of the most visited and commented communities in the LJ-based blogs, namely *Oh no, they didn't!*, which claims, in its headline: "Celebrities are disposable. Gossip is priceless" to summarize its agenda. Fandom and gossip are a major blogging activity. Bloggers are huge fans of some celebrities and write about, comment on and defend their idols in passionate terms. *Oh no, they didn't!* displays several sub-communities. They are listed in Table 3.1 and can be grouped around a sub-genre (e.g., a TV show, such as *Star Trek* or *Glee*) or around a geographical community (e.g., *Arama they didn't!*, about "Japanese Entertainment News," and *Omona they didn't!*, about "Korean Pop Culture"). Connected to "Fandom and gossip," the growing phenomenon of "Fanfiction" represents another significant trend in current blogging communities, even though it is a relatively recent emerging fictional genre. It is one of the most restricted categories as far as access to non-fans is concerned and implies writing entries, or more sustained pieces of writing, involving characters and settings related to their idols, be they characters from books, movies, cartoons, TV shows, and so on. Such bloggers usually take for granted that their readers are very well familiar with the stories and characters they write about and, hence, write niche fiction that can be appreciated and in some cases fully understood only by acolytes.

One of the most popular LJ blogs, called *Fandomsecrets*, is based on anonymous entries sent by fans, who reveal a secret about their fandom and are regularly published in series. What is interesting about this community is the amount of integration of images and writing, as each entry is made up of an edited image, featuring characters, TV/movie shots or other fandom-related materials, plus written captions inside the image(s). For example, one post/image featuring three pictures of the character from the TV show *The Walking Dead*, embedded in its caption, thus forming an ensemble of words and images. The entry is partially reproduced in Figure 3.1.

The caption reveals a very common pattern that is evident in many fandom blogs and communities, that is a contrast between everyday life and the psychological interference that their media/digital life exerts on them. The mixing of the two planes, that is, real life (i.e., marriage) and an intimate, fandom-related life, can be fully appreciated through the de-coupling and interpretation of the two semiotic resources, that is, image plus caption. From a multimodal point of view, however, what is more interesting to note is the complete integration of the two semiotic resources used to produce the entry. In apparent contrast to findings in section 3.3, this entry (and all the entries featured in this blog/community) perfectly integrates writing and image, ultimately erasing the very notion of resource-switching, as no switch is used here, but a complete embedding of writing into images.

Another anonymous entry from this community is partially reproduced in Figure 3.2. The integration of image plus caption is total. The image has been modified for copyright reasons, but it keeps the exact layout and wording used in the original.

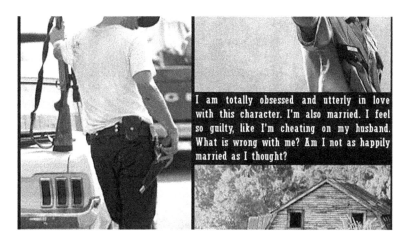

Figure 3.1   Fandomsecrets' entry

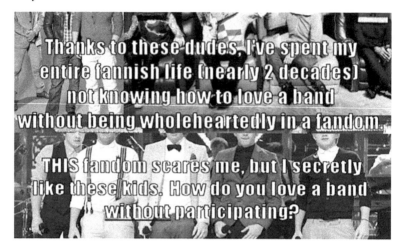

Figure 3.2   Fandomsecrets' entry

This example highlights another seminal aspect in these communities: (1) loving a band (or any other kind of passion) makes sense *only thorough fandom*; and (2) the essence of fandom is *participation*. This sense is reinforced by sharing secrets that seem impossible to reveal if not covered by anonymity that is paramount in such a community.

Fanfiction is in effect a textual experience that cannot be ascribed solely to blogging practices, as it cuts across blogs, fan communities and social networks, which are embedded one into another. An entry from the blog reports, entitled "How authors feel about fanfiction," is worth being partially reproduced here:

George R.R. Martin.

Martin has always been against fan-fiction, which must be a little sad for the fan-fiction writers out there, seeing as how very many delicious characters he has created. He writes, "Every writer needs to learn to create his own characters, worlds, and settings. Using someone else's world is the lazy way out." But more importantly, in a very articulate and informative explanation on the legal and monetary problems with fanfiction, he explains, "My characters are my children . . . I don't want people making off with them, thank you. Even people who say they love my children. I'm sure that's true, I don't doubt the sincerity of the affection, but still . . . No one gets to abuse the people of Westeros but me."

J.K. Rowling.

The author of the beloved *Harry Potter* series takes a more amenable view to fan-fiction, and though she has the same commercial issues as other authors, she's basically given it her blessing. Unlike Martin (who, let's face it, wouldn't have a leg to stand on with this argument), she seems to be most concerned that any Potter fan-fiction remain PG-rated. According to an official statement from her agent, "[S]he is very flattered by the fact there is such great interest in her Harry Potter series and that people take the time to write their own stories. Her concern would be to make sure that it remains a non-commercial activity to ensure fans are not exploited and it is not being published in the strict sense of traditional print publishing . . . The books may be getting older, but they are still aimed at young children. If young children were to stumble on Harry Potter in an X-rated story, that would be a problem." (Goofusgallant 2012)

As can be seen from the post above, the question of fanfiction is catching the attention of the general public and also of creative writers and practitioners in the field.

Another broader category is devoted to sport and leisure, also including hobbies and interests, but *excluding* anything that goes into the two former categories, namely fandom related to TV shows, cinema, music, cartoons and video/digital games. This category usually implies personal participation in practiced activities, such as sports or cooking. Another context-specific category is devoted to the development of digital community-related tools, for example tutorials, thus explaining how to create an animated gif for example, or posting icons to be used by other bloggers. Self-help has been separated by the more general "asking for information" category, but is eponymous to it and may be also described as an example of blegging.

3.3   A MULTIMODAL ANALYSIS: ISSUES AND DEFINITIONS

The linguistic analysis is here complemented by some observations on a purposely-created corpus, the LJ corpus including other semiotic resources (LJC_SR), which consists of several semiotic resources listed in Table 3.2.

As can be seen from Table 3.2, the resources can be grouped under three main headings: (1) language (mainly written), (2) images and (3) videos.

Verbal language is mainly written, because the genre analysed in this chapter is the *written blog*. Vlogs have here been excluded, but the videos are sometimes included in the ensemble of resources used in the blogging community. Spoken language thus mainly refers to its use in video clips that bloggers attach to their entries, incorporating music and sound. A typical configuration of resources is reported in the following representation. This representation shows that the resources are embedded, but hold significantly different positions: in the centre, the independent variable is the website template, that is, the general layout where entries are embedded. Another typical feature is the blogger's avatar. Some optional personal

Table 3.2   Semiotic resources in LJC

Verbal written language
Verbal spoken language
General layout (i.e., website template)
Images

  Pictures
  Drawings
  Photos
  Avatars
  Animated gifs
Videos

  Music
  Sound, background noise

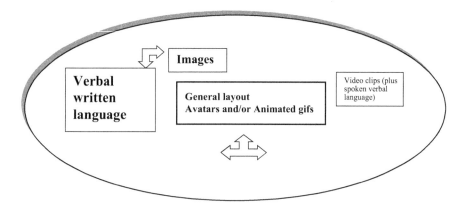

information, such as email address and gender, are usually signalled by the use of specific icons (e.g., colour for gender) and/or animated gifs. The latter is a file made up of an animated graphic image on a web page, for example presenting in sequence a small scene featuring famous actors or singers, or nature scenes that the blogger may wish to use as a personal sign of identification within the community. These can loop *ad libitum* or stop after the animated sequence has been played once. The blog's template and bloggers' identification clues, which may vary from a bare minimum to a wealth of information, are often given by means of graphic or iconic strategies. These are placed in the centre of the representation, as they signal the core of multimodal blogging practices. These resources are affordances, in that they represent technological *and* semiotic constraints at the same time.

Written discourse, the most representative and used resource in traditional blog, is placed on the left. As several studies have claimed, writing is still predominant in most digital environments. However, images are also significant and have been found very consistently across different blogging communities. Be they in the form of pictures, drawings or screenshots, they are very frequently used. Bloggers are also keen on showing their proficiency with both animation and graphics editing programmes, for example creating their own animated gifs, customized icons and pictures. This is also attested by the significant incidence of tutorials within blogging communities, that accounts for about 14 percent in more than 3,700 of the most visited blogs. Whatever the case, images make up the second most important and frequently used resource within blogging communities, as they are used for a variety of communicative purposes. Incidentally, we may note in passing that professional and corporate blogs also highly employ visual resources, for obvious reasons, because they are much cheaper in the digital industry than in the printing industry.

Also note that the arrows from the *verbal written language* box and the *images* box are reciprocal, and the same may be said about the general genre affordances and the *verbal written language* box and the *videos* box. This is not the case with the *images* box and the *videos* box, where the lack of arrows represents the near-absence in the association of the two different sets of resources. This is also consistent with the different size and highlighting of the different boxes and resources. On the right are the less frequently used resources of videos and video clips. Vlogs, as I have said, are made up of videos, whereas "traditional" blogs use videos less than might be expected. Randomly sampled blogs from *LiveJournal*, *Blogpress* and *Wordpress* feature a total of 3–4 percent of videos in personal blogs and blogging communities.

LJC is summarized in Table 3.3.

LJC has been divided into ten small subsections, including 150 entries each, named LJC1, LJC2, and so on. This initial operation was carried out to saturate the corpus from a lexical point of view. Furthermore, this has been done to facilitate manual operations on smaller datasets. See the Table 3.4 and 3.5 to compare LJC1–LJC10.

3.3.1   Analysis of Semiotic Resources and Reciprocal Interactions in the Corpus

Despite possible benefits, comments have not been included in LJC, as it would have been arbitrary to set a time limit for the insertion of comments in the corpus. Blog entries are relatively stable and most of them are permalinks, whereas comments, with related questions of visibility, are generally

Table 3.3   LJC in numbers

Blogs	Entries	Word number	Images	Tags	Videos	Entry's title	No entry's title	Author's nickname	Animated gifs	Comments
329	1500	926120	5385	3149	235	1298	202	1459	333	132007

Table 3.4   LJC in subsets: LJC1–LJC5

	LJC1	LJC2	LJC3	LJC4	LJC5
Entries	150	150	150	150	150
Word number	105512	54804	88745	134780	78945
Images	365	609	435	403	497
Tags	106	243	305	479	700
Videos	19	33	12	23	53
Entry's title	134	126	125	112	114
No entry's title	16	24	25	38	36
Author's nickname	142	150	140	147	150
Animated gifs	14	12	34	32	78
Comments	11479	15762	12789	18734	10890

Table 3.5   LJC in subsets: LJC6–LJC10

	LJC6	LJC7	LJC8	LJC9	LJC10
Entries	150	150	150	150	150
Word number	114894	97872	67890	123006	59672
Images	540	756	768	485	527
Tags	129	331	438	179	239
Videos	18	23	32	8	14
Entry's title	138	121	139	142	147
No entry's title	12	29	11	8	3
Author's nickname	145	147	145	150	143
Animated gifs	23	32	54	16	38
Comments	13450	11479	15762	11792	9870

less easy to define. In particular, the extreme rapidity of change in comment number for popular entries in *LiveJournal* would complicate any operation of inclusion and exclusion, as they increase in number very quickly and would change while the corpus is being constructed. Corpus size has been designed according to entry number, as entries were selected as sampling units. As Tables 3.4 and 3.5 show, the corpus has also been divided into smaller datasets, containing 150 entries each. As a consequence, word number varies to a certain extent, as is not possible to predict or set the word number for each entry in advance. However, this method has been used to handle smaller datasets.

If we consider written verbal language and, thus, the number of words and the number of comments together, it becomes apparent just how much more often this resource is used in comparison to any other in this context. Incidentally, note that in Tables 3.4 and 3.5, the number of comments themselves rather than the number of words used in the comments has been incorporated. Writing is clearly used far more than any other resource, as it is extensively used in overall LJC and the word number/other resource percentage is as follows: (1) image/word number 0.59 percent; (2) tag/word number 0.35 percent; and (3) video/word number 0.025 percent. Overall word mean (average) in the corpus is 92.612, whereas the median is 93.308.

The resources interacting with the written text (i.e., entries) may be classified according to several criteria that cannot be determined without some degree of ambiguity. However, images may represent different kinds of visual statements and the same may be said about videos, that are also audiovisual statements accompanying written verbal entries (e.g., captions).

A different role is instead played by the practice of tagging. A tag is a keyword, in the form of metadata, which allows different kinds of associations. A text can be thus lexically associated with another text, an image can be associated with another image, but they can be also grouped semantically, crisscrossing resources and giving rise to an intersemiotic semantic network of relationships. Tagging thus represents an interesting trait that is both linguistic and non-linguistic at the same time. Tags are made up of words and, as such, are constituted by verbal language, but they also refer to other texts or images or videos, playing the role of *decontextualized textual cues*.

Textual displacement may also be interpreted as a practice of ordering the chaos of millions of digital texts that are difficult to access, make sense of and be assessed by users/readers. If a blog is a macro-text and *Twitter* posts are micro-texts, tags are *hyper-micro-texts*. They can consist of an isolated word, but also comprehend more than one word and give indications as to what the tagger has categorized and also other information for future users. Furthermore, tagging may be used in different ways: it can be highly personal, giving information about the tagger's opinion on what is tagged, thus failing to provide truly useful semantic associations, replicable and interpretable univocally, or at least partially univocally.

With regard to other information, the name and basic information (e.g., gender) are included in the corpus. 96 percent of bloggers include a nickname in their posts, 86 percent of entries are titled and only 14 percent do not display a title. The system of giving titles to one's own posts may be also interpreted as a preferential strategy in blogging communities, along with the use of the author's nickname, which is required in most platforms.

The interaction of semiotic resources in these environments is far from straightforward. The almost ubiquitous presence of a personal publishing platform impinges on the old-fashioned idea that blogs are free and far from corporate influences and constraints. This study, as much of current research literature, rejects the view of such unbridled textual practices. Technical affordances play their role in shaping texts and they should be taken into account when assessing communicative practices, let alone digital communicative practices, that are conditioned by the virtual setting in which they take place.

3.3.2   Resource-switching

A first analysis of LJC seems to suggest a negligible presence of non-verbal resources within blogging communities. The low degree of use of non-verbal resources may seem puzzling at first, and especially so if compared with any first-hand—even cursory—experience with web-based environments. How is that images amount to an almost insignificant 0.59 percent when compared to the written word? Are we still living in a logocentric age?

Some considerations are paramount for a better understanding of such findings. First of all, posts are usually more based *either* on writing *or* on images, more rarely on both of them in the same entry. There is a limit to the number of words and images in most blog communities. For practical reasons, a screen contains far more words than images (as a printed page does). Images may be small or compressed, but *one image is usually a larger text than a written text*. It is certainly not merely a matter of actual size or of a portion of the screen, but it is nonetheless true that space and layout are medium-constrained features in CMC. It may be consequently assumed that words and images cannot be equated, but need to be computed according to different criteria.

For a better appreciation of the percentages of occurrence between image/word, tag/word and video/word, we need to observe this phenomenon from a closer perspective, singling out smaller datasets, also considering that a manual and semi-manual analysis is better suited for a closer inspection of the phenomenon that will be labelled henceforth as *resource-switching*.

As said, resource-switching is different from mode-switching in that the former involves a wider range of resources, including verbal language *tout court*, but excluding those delicacies such as the difference between verbal spoken language and verbal written language, as the latter does. However, both resource-switching and mode-switching imply any kind of *alternation* in the same digital communicative event.

Why is it important to study these alternations? Why is it important to understand why, when and to what extent participants make use of these alternation opportunities in a web-based environment? Take this example: if in a face-to-face communicative exchange, a speaker first spoke and then switched to writing or scribbling something on a slip of paper and then passed it to one other specific speaker, this would be *a marked choice*. If a speaker took a photo of one of his/her interlocutors during a conversation, this would be certainly significant and convey a meaning external to the situation in hand. It is likely that this would not go unnoticed.

However, when we come to digital environments, we seem more than willing to accept a range of communicative choices among groups of people without questioning them. Today, we are so inundated or even swamped by communicative resources and so used to making full use of them, changing, juxtaposing and alternating them, that it seems almost pointless to investigate the patterns of such rapid, taken-for-granted and apparently unpredictable mixes. It is "natural" that during a Skype video conversation participants use both speech and writing at the same time. When questioned for the book survey, interviewees appeared more than puzzled when they had to reflect on these alternations between voice and text, even though these patterns were used extensively and regularly. It is also quite normal and typical for bloggers to use pictures, wording, videos and many other resources in their blogs. It would be a very marked choice if they did not!

The examples made above from daily contexts are contrived, but they have been presented to show that digital practices are hardly comparable to everyday live interaction, be it mundane or not. Synchronicity is one of the determining factors that distances live interaction with *some* (most, but not all) web-based communicative events. Synchronicity alone, however, cannot explain all the semiotic, technical and linguistic distinctions between face-to-face and web-based interactions. Other features are important, such as the technical and semiotic affordances of the two distinct communicative situation types (*live* vs. *virtual*).

As far as the latter is concerned, it cannot be overstated that material aspects affect how communication takes place. The fact that a set (whether small or big) of resources are available for communication influences the decision to use them. To put the matter in simple terms, if resources are available, why not use them? Furthermore, communication is consequently augmented and nourished to hypertrophy by these resources, which have been investigated by several strands of multimodal studies.

Literature in the field of multimodal studies is wide and varied, and presents the study of each of these resources from several angles and standpoints. For example, the resources of image and colour (Kress and van Leeuwen 1996, 2002, 2006), gesture and movement (Kress et al. 2001, 2004; Martinec 2004), body movements, social distance and gaze (Harrigan 2005; Sindoni 2010, 2011a, 2011b, 2012), voice and music (van Leeuwen 1999), and space (O'Toole 2010). Multimodal studies have focused on a variety of different objects of study, such as visual design (van Leeuwen and Jewitt 2001), interactional practices (Norris 2004), production and distribution of objects or places (Kress and van Leeuwen 2001), educational settings, practices and mediascapes (Marsh 2005), or the production of digital texts, film, adverts, web pages, mobile culture, movie language and digital communities (O'Halloran 2004; Scollon and LeVine 2004; Leander and Vasudevan 2009; Piazza, Bednarek, and Rossi 2011).

As these and other studies show, resources may be highly integrated in digital environments, but the fact that they are there does not fully explain or justify such extensive use. However, the above claimed hypertrophic use seems disconfirmed by this study. Even a basic familiarity with the web-based environments discussed in this book may suggest that maybe things are much more complex than these initial findings seem to imply. Furthermore, the presence and consequent use of a wealth of resources does not necessarily explain their effective use.

For example, material and technical affordances are important to shed further light on these issues. Technical aspects also include, at an even more basic level, the use of a screen and a web connection, which are necessary for any web communicative event. The screen, as discussed in the previous chapter, is the virtual arena where communication shapes web interaction and is shaped by it. The screen is, in other words, a technical constraint bestowing semiotic affordances. It is, broadly speaking, a framework, a setting where communication takes place: to put the matter in concise terms, we shape our conversations in "screeney" terms, to borrow from Baldry and Thibault (2006). Conversely, when we are in a face-to-face situation, we are influenced by other factors, such as the setting shared by participants. Synchronicity alone is not sufficient to account for differences concerning the material or physical conditions of live vs. virtual interactions.

With regard to the semiotic resources deployed in the two different environments, it would be useful to recall the discussion in Chapter 1 at this point. Some resources can be used in any kind of environment, setting or situation type, be they real or virtual. However, if they are strictly connected with our body physical semiosis, they are severely modified by the digital environment: think of how social distance (i.e., proxemics) or gaze are altered in web-based video interactions.

Other resources, such as verbal language and images, can be used in both real and virtual settings, but they too are significantly altered according to the situation. However, as explained in the Introduction, verbal language, be it written or spoken, is camouflaged in digital interaction, in the sense that it is apparently much alike what it is in a face-to-face context.

Words are always words, aren't they? The answer that this book is providing is *no*. Words are turned into something profoundly different, even though they "look pretty much the same." The use of body resources (e.g., body movements, gaze, social distance, etc.) is easily recognizable as different in digital interactions if compared to face-to-face encounters. Of course, we cannot shake hands and reciprocate gaze as if we were in the same place. We cannot smell nor slap someone on the back on the web.

Images are also quite different in face-to-face and digital interactions: we can send pictures, post images, and fill our blogs with drawings or photos, communicating with them non-verbally, but the same situation is hardly replicable in a live context, with some exceptions, for example with art exhibitions. The difference in semiosis, production and use is evident, out there to be perceived. The same cannot be said about verbal language. Words are always words . . . aren't they?

This book suggests that words seem the same, but in fact are not. In the following section, resource-switching will be used as a paradigm to explore patterns of alternation.

3.3.3 Resource-switching in Context

Corpora LJC1 and LJC2 have been selected as they present a considerable contrast in the image-word ratio, that make them a useful comparison for heuristic purposes. LJC1 presents 105,512 words and 365 images, whereas LJC2 presents 54,804 words and 609 images. LJC1 and LJC2 present an average of 80,158 for words and 487 for images. Image/word percentage is 0.35 percent for LJC1 (word/image ratio: 289.07) and 1.11 percent for LJC2 (word/image ratio: 89.99).

In LJC1, only 51 entries out of 150 included images ($\geq$1). In these 51 posts, the post/image ratio is 7.15, with a percentage amounting to 13.9. In LJC2, only 60 entries out of 150 entries included images ($\geq$1). In these 60 posts, the post/image ratio is 10.15, with a percentage amounting to about 9.85. A more delicate categorization is useful, thus separating the blogs containing images (henceforth, *posts-cum-images*) from the blogs not containing images (henceforth, *image-free posts*) and creating two other subcorpora for LJC1 and LJC2, each of them gathering only posts with images, but deleting them to count *only word number*. These two subcorpora of *posts-cum-images* have been called, respectively, LJC1_Images and LJC2_Images, the former including 51, the latter including 60 entries.

In LJC1_Images, only 14,996 words were counted from the previously selected 51 posts, including the same grand total of 365 images: the word/image ratio is about 41.08, thus significantly altering the previous findings, whereas LJC1 combines both *image-free posts* and *posts-cum-images*. The percentage of images per word here rises to 2.43 percent. It is also interesting to note that almost one-third of LJC1 (i.e,. 51 posts) amounts only to 14,996 words (14.21% of the total), whereas the remaining part of LJC1, including *image-free posts*, amounts to a striking 90,516 words (85.78% of the total). This indicates that *posts-cum-images are less wordy than image-free posts* that, conversely, feature a significantly higher word number. In other words, *there is a significant disproportion in word number between posts-cum-images and image-free posts.*

LJC2_Images contains 60 *posts-cum-images*, including, respectively, 19,848 words and 609 images, for a word-image ratio of 32.59. The percentage of images amounts to 3.06 percent, rising in comparison to LJC1_Images. LJC2 includes 60 *posts-cum-images* and their word number amounts to 36.21 percent of the overall LJC2 word number. The *image-free posts* amount to 34,956 word number (63.78%) of the whole LJC2, whereas the *posts-cum-images* amount to 19,848 word number (36.21%

of the total). Here, the disproportion is less striking, but it nonetheless confirms that which resulted from LJC1 and the exploration into the subcorpus LJC1_Images: the *posts-cum-comments* are significantly wordier than *image-free posts*.

The overview of these findings in Table 3.6 summarizes aspects treated so far in *posts-cum-images* subcorpora LJC1_Images and LJC2_Images.

These findings may lend themselves to several interpretations, even though it cannot be overstated that such corpora are too small to allow for generalization. However, some heuristic observations may guide future research on the matter. The purportedly high integrability of semiotic resources on the web, and especially so on Web 2.0, needs to be carefully considered according to web genres and texts. Macro-genres, such as web pages and social networking communities, such as *Facebook* and *Reddit*, and video sharing communities, such as *YouTube*, display a huge number of semiotic resources, such as words, images, videos, music and so on. We all, as internet users, are familiar with digital environments, where many resources are deployed at the same time to convey meanings, multiplying levels of understanding and creating different reading pathways, as research literature has attested (Sindoni 2011b). However, it is important to understand that the above-mentioned digital practices are embedded in *corporate digital environments*. A high integrability of resources, even the mere co-deployment of several resources, is the product of careful, planned and purpose-driven design, that is, in one way or another, aiming at profit, as the obvious case of advertising shows.

In less-constrained digital environments, at least when users take the lead in shaping their own texts and in choosing the resources through which communication is instantiated, it may be assumed that there could be less balance in resource use. These findings seem to suggest that users, when creating their own texts (blog entries) in blogging communities, favour one resource over another, for example with reference to writing and image.

Generally speaking, writing is much more used in the overall LJC, but when images are preferred, writing plays a secondary role. In manual qualitative observations of *posts-cum-images*, it has been noticed that writing is largely subservient to images, as it is employed to describe them, mainly in the form of captions.

Table 3.6 Data on *posts-cum-images*

	Entries with images	Word-image ratio	Percentage of images per word
LJC1_Images	51	41.08	2.43%
LJC2_Images	60	32.59	3.06%

3.4   BLOGGING AS ONGOING CONVERSATION: A CORPUS LINGUISTICS CASE STUDY

The aim of this section is to explore language variation using corpus linguistics methodology to extract data from our datasets in LJC. This chapter thus makes use of keyword and cluster analyses (Scott and Tribble 2006) and occasionally draws from MF/MD analysis, placing positive and negative loadings of features along the dimensions of the spoken/written variation in text genres as identified by Biber (1988, 2003). Using Wordsmith 5 and 6 (Scott 2008, 2012), a keyness analysis was carried out to see the most prominent linguistic features that are functionally related to the blog as a text genre. Keywords are the words whose frequency is unusually high (positive keywords) or low (negative keywords) in comparison to a reference corpus. Keyness is defined in terms of statistical unusuality. A key cluster is similar to a key word, but concerns longer chunks or syntagmatic structures rather than single words and may be revealing in addressing issues of text genre in terms of the *aboutness* and also in further explorations concerning spoken/written variation.

Lexical bundles have also been identified to explore relevant questions, mainly concerned with variation in the spoken/written *continuum*. This investigation is aimed at showing whether lexical bundles are frequent and the extent to which they vary across spoken/written genres. A multifeature/multidimensional approach to variation can be a powerful tool for exploring variation across genres, but studying variation through lexical bundles can approximate an MF/MD analysis (Tribble 1999).

Variation has been computed modifying some standard default settings of Wordsmith, that is a minimum frequency of 10 and a maximum of 16,000 (which is large enough to include all keywords), with Dunning's Log Likelihood statistics with a *p* value of 0.0000001 and full lemma processing. As a preliminary step, word length and a standardized type/token ratio have been computed using the Wordlist function of Wordsmith for LJC, as is detailed in Table 3.7, which also reports on other significant data.

Note that type/token ratio has been standardized, as all frequencies have been normalized to a common 1,000-word basis to guarantee comparability between files of different size in the corpora taken as reference. Five wordlists have been created, using three different corpora: (1) the one-million word ICE GB corpus (overall), (2) the ICE spoken component, (3) the ICE written component, (4) the one-million word FLOB corpus and (5) the 100-million word BNC World corpus.

The ICE GB corpus (*International Corpus of English, Great Britain*) includes 200 written and 300 spoken texts, and has also been split into two subcorpora to explore spoken/written variation through a subsequent keyness analysis. The other two selected corpora are the one-million FLOB corpus and the 100-million BNC World corpus. The FLOB corpus is the Freiburg-LOB Corpus of British English, an update of the LOB corpus of

Table 3.7   Data on LJC

text file	LJC_.txt
file size	1,144,234
tokens (running words) in text	193,723
tokens used for word list	191,172
types (distinct words)	12,299
type/token ratio (TTR)	6.43
standardized TTR	43.96
standardized TTR standard deviation	54.95
standardized TTR basis	1,000
mean word length	4.34
word length standard deviation	2.28

the early 1990s, and includes only written texts, including *news, fiction* and *academic prose*, whereas the second is the British National Corpus–World Edition, which includes both spoken and written texts, but with a prevalence of written texts.

A preliminary analysis of the top 100 items of the four wordlists, sorted by frequency, shows that the most frequent content words in LJC are: *said, like, know, time, get, want, see, comments, think, go, http, man* and *make*; whereas in the overall ICE wordlist they are: *know, think, like, people, time, mean*; in the FLOB wordlist: *said, new, time, like, people, made, years*; and in the BNC World wordlist: *said, time, like, new, know, people, see, get*.

What emerges is that the LJC wordlist presents a higher number of content words and, if we except the acronym *http* and the verb *get*, a consistent occurrence of private verbs, such as *know, see* and *think*, expressing intellectual and non-observable states of mind and public verbs, such as *say*, implying observable actions, mainly used to introduce indirect statements. In LJC, FLOB and BNC World wordlists, *said* is the first lexical verb in either the simple past or past participle form. With the exception of *said*, all other verbs appear in their base form in the LJC wordlist: *like, know, want, see, think* and *make*, all (with the exception of *make*) implying a perceptive, cognitive, desiderative and emotive mental process (cf. Halliday and Matthiessen 2004). Some of these frequent mental verbs are also found in the other corpora wordlists.

The LJC wordlist also presents a higher level of *aboutness* that may be referred to the specific community of blogEng, involved in *commenting* and *liking* activities, as suggested by occurrences such as *comments* and *like* that are likely to refer to common practices in digital environments. They point, in fact, to interactional communicative strategies that allow bloggers to act on the activities of other bloggers or users, either by commenting on them or simply by expressing explicit agreement (i.e., *like*).

A second step was taken to ensure that the size and genres included in terms of the spoken-written variation of the reference corpus were reliable. Some additional wordlists were used to act as reference corpora for LJC. First of all, keyword lists created using the three reference corpora were compared for both positive and negative keyness. The three corpora used as reference were selected for their size difference. FLOB and ICE are similar in size, whereas BNC World is larger, but also because of their significant difference with regard to the inclusion of spoken/written data, as FLOB is entirely made up of written data, whereas ICE and BNC World include both spoken and written data, albeit in different proportions. Furthermore, these three corpora encompass a wide range of varieties of the English language (see the corpora websites for further details on the reference corpora design).

Table 3.8 shows the top ten keywords in the selected three reference corpora (RC).

The three keyword lists, sorted for positive keyness, display eight positive keywords in common: the FLOB and ICE-based lists display nine common positive keywords, the FLOB and BNC World, nine common positive keywords and BNC World and ICE, eight common positive keywords. However, the order in which they appear is different. Many items appear as positive keywords in all three lists, due to inconsistent transcription and evidence of this is in the presence of items such as *you* (with or without enclitics), *I'm*, *it's*. Generally speaking, the positive keywords are, as is easy to expect, a number of content words, such as personal nouns, toponyms and "digitally-related" items, such as *comments, http, www, com, tags*. As with the negative keywords, see Table 3.9.

With regard to negative keywords, four items are common to the three keyword lists and they appear in different word order. However, if we

Table 3.8 Top ten positive keywords with three different reference corpora

No.	FLOB as RC	BNC World as RC	ICE GB as RC
1.	you	Jin	his
2.	Jin	Kazuya	he
3.	it's	http	Jin
4.	Kazuya	it's	him
5.	his	his	Kazuya
6.	I'm	comments	it's
7.	http	he	I'm
8.	he	him	her
9.	comments	I	http
10.	him	you	comments

Table 3.9   Top ten negative keywords with three different reference corpora

No.	FLOB as RC	BNC World as RC	ICE as RC
1.	of	the	uh
2.	the	of	uhm
3.	which	in	s
4.	in	which	of
5.	s	by	ve
6.	by	er	is
7.	its	its	the
8.	is	is	yeah
9.	government	government	yes
10.	such	per	which

compare the FLOB and BNC World-based lists, the common negative keywords rise to eight items, the BNC World and ICE-based lists have four
common items and the FLOB and ICE-based lists, five common negative
keywords. *Uh* and *uhm* appear as the first and second top negative keywords in the ICE-based list and they are both typical of spoken conversation. Biber does not consider interjections as discourse markers (1988),
but Schiffrin (1987) does, as she attaches them with the value of keeping
conversational coherence.

As is evident from a rapid comparison, the FLOB and BNC World-based
lists are much more similar than the ICE-based list. In ICE, the prepositions
in and *by* are in sixteenth and twenty-eighth position, respectively. If we
also take into account the inconsistent transcriptions of items in contracted
forms, such as *'s* and *'ve*, the differences are less striking.

However, these differences are nonetheless noteworthy: why is it that the
FLOB-based and BNC World-based lists are more similar, whereas the ICE-
based list presents different negative features? A partial answer may lie in
the fact that the first two reference corpora are based more on written texts:
to be more precise, the FLOB corpus incorporates exclusively written data,
whereas the BNC World is mainly based on written texts. As a consequence,
typical features of written texts appear as negative keywords in that they are
less prominent in LJC, that is also *exclusively* based on written data, but of
a specific nature.

The first two keywords in FLOB and BNC World are *the* and *of* (even
though in a different order). *Of* as a preposition adds a negative weight
to Dimension 1 (i.e., involved vs. informational production), identified by
Biber's MF/MD analysis of the spoken/written variation in genre analysis
(1988). Tribble (1999) claims that *of* and *the* are usually associated with
nouns, also observing that in academic prose *of* is used as a postmodifier
in the N1 + *of* + N2 structure. The definite article *the* is also associated
with nouns and in Biber's MF/MD analysis, nouns of the nominalization

type (i.e., *government* in both FLOB and BNC World) are a feature with a positive loading for Dimension 3 (i.e., explicit vs. situation-dependent reference), while nouns of other types are a feature with a negative loading for Dimension 1 (i.e., involved vs. informational production). These features thus point to the negative prominence of such items in the LJC-based list, suggesting that some items commonly associated with written prose are less prominent than could be expected in a corpus based on written data.

However, a comparison with the ICE-based list sheds light on another aspect: four of the top ten negative keywords are typical of spoken conversation. Interjections such as *uh*, *uhm*, *yes* and *yeah* have positive loading along Dimension 1 (involved vs. informational production). Similarly, contractions such as *'s* and *'ve* may be considered as positive indicators along Dimension 1 as well, even though the issue of inconsistent transcription needs to be taken into careful account and may cause incorrect interpretations with regard to enclitics. The ICE-based list, however, presents mixed features: for example *which*, an item that is also present in the two other negative keyword lists and displays positive loading along Dimension 3 (explicit vs. situation-dependent reference), being an indicator of either WH-relative clauses and pied-piping constructs.

The fact that the two keyword lists generated with two different corpora yielded similar results suggests that corpora size and other variables did not affect the outcomes, to any considerable extent, despite some significant differences.

3.4.1   A Keyness Analysis: BlogEng

After the initial steps discussed in the previous section, aimed at compiling wordlists and also keyword lists based on different reference corpora, the second phase of this exploration of blogEng, that is the English language used in blogs, was undertaken through the selection of one reference corpus for in-depth analysis.

The reference corpus in question is the one-million word ICE GB corpus (see Appendix 3.1 for corpus details), which will be also segmented into two separate subcomponents in this section, namely the ICE spoken and ICE written subcorpora. Four separate wordlists were used for this purpose: (1) LJC1-2, whose statistical details were reported in Table 3.7; (2) the overall ICE (i.e., ICE_all); (3) the ICE spoken (i.e., ICE_spoken); and (4) the ICE written (ICE_written). These wordlists were subsequently employed to create three new keyword lists, comparing LJC_all with the three different corpora-based reference lists, namely ICE_all, ICE_spoken and ICE_written, but this time changing settings for Wordsmith, as to reduce both maximum wanted keywords to 500 and to increase minimum frequency to 10.

Tables 3.10 and 3.11 show, respectively, positive and negative keywords for the three reference corpora.

Table 3.10   Top ten positive keywords with three different reference corpora

No.	ICE as RC	ICE_spoken as RC	ICE_written as RC
1.	his	his	his
2.	he	Jin	he
3.	Jin	he	I
4.	him	her	him
5.	Kazuya	him	Jin
6.	it's	it's	you
7.	I'm	Kazuya	Kazuya
8.	her	I'm	it's
9.	http	comments	I'm
10.	comments	http	her

Table 3.11   Top ten negative keywords with three different reference corpora

No.	ICE as RC	ICE_spoken as RC	ICE_written as RC
1.	uh	uh	the
2.	uhm	uhm	of
3.	s	s	is
4.	of	I	which
5.	ve	that	by
6.	is	yeah	be
7.	the	yes	in
8.	yeah	ve	may
9.	yes	it	are
10.	which	well	however

Comparing the data, the top ten items in the three ICE-based lists have eight positive keywords in common, all (i.e., ten) between the general ICE-based list and the spoken ICE-based list, eight between the overall ICE and the ICE _written, and eight between the spoken and the written ICE-based lists.

With regard to negative keywords, the picture is even more revealing. The overall ICE-based list has six negative keywords in common with the spoken ICE-based list, and four negative keywords in common with the written ICE-based list. Consistent with this finding, the ICE-based spoken and the written ICE-based keyword lists display a remarkable difference in the most prominent negative keywords, as they do not have a single item in common.

Such findings suggest that LJC lies somewhat in-between the spoken and written genre, as it seems to lack some of the typical features of spoken genres, for example interjections such as *uh, uhm, yeah, yes,* and discourse markers such as *well.* In a similar fashion, *that*-clauses as verb complements, *that*-relative clauses and *that*-clauses as adjective complements have positive loading along Dimension 6 (i.e., online informational elaboration).

The lack of such features would thus suggest the prominence of more written-related features, but the top negative keywords in the written ICE-based list show a similar pattern in that typical features of written genres are also absent from LJC-based keyword list. *The* and *of* have already been discussed, but other features also signal significant variation toward the written mode. For example, *which* for sentence relative has positive loading along Dimension 1 (i.e., involved vs. informational production), *by* may signal by-passives and together with the conjunct *however* display positive loading along Dimension 5 (i.e., abstract vs. non-abstract information). However, the prepositions *by* and *in* have negative loading along Dimension 1 (i.e., informational vs. involved production).

To sum up these findings, we may argue that LJC includes highly hybridized and polymorphic texts. They are all written entries, but show some features that may be associated with speech. However, claiming that they are "spoken-like" texts would be far-fetched, in so far as they also lack typical markers of spoken interaction, such as interjections or discourse markers. However, they also lack significant features related to typically written genres.

An analysis of positive keyness in the general ICE-based list reveals the results shown in Figure 3.3.

If we ignore the personal nouns, such as *Jin*, *Kazuya*, *Jack* and *Jeff* (all coming from fanfiction stories), a considerable amount of unexpected high frequency of personal pronouns (e.g., third person, first person) plus enclitics *'s* and *'m* appears. With regard to personal nouns, a double layer of provenance becomes evident. Jin and Kazuya clearly come from Eastern/Asian countries, whereas Jeff and Jack could not be more Anglo-American.

Keyness values, appearing in the sixth column, are particularly high. This high degree of occurrence seems to suggest a high level of characterization, that may be found especially in fanfiction. If the first three items are considered (ignoring the personal noun *Jin*), *his*, *he* and *him* appear, hinting at a particularly masculine world. *Her* is eighth for keyness, thus suggesting

N	Key word	Freq.	%	RC. Freq.	RC. %	Keyness	P	Lemmas	Set
1	HIS	2,288	1.18	1,091	0.20	2,503.51	0.00000000		
2	HE	2,731	1.41	2,633	0.49	1,445.19	0.00000000		
3	JIN	468	0.24	0		1,241.86	0.00000000		
4	HIM	1,038	0.54	545	0.10	1,053.28	0.00000000		
5	KAZUYA	354	0.18	0		939.20	0.00000000		
6	IT'S	357	0.18	2		923.65	0.00000000		
7	I'M	327	0.17	0		867.54	0.00000000		
8	HER	1,085	0.56	781	0.15	824.67	0.00000000		
9	HTTP	277	0.14	0		734.83	0.00000000		
10	COMMENTS	309	0.16	17		696.72	0.00000000		
11	JACK	282	0.15	9		673.36	0.00000000		
12	WWW	213	0.11	0		565.00	0.00000000		
13	JEFF	208	0.11	1		539.67	0.00000000		

Figure 3.3 Keywords sorted by positive keyness

a lower level, at least from this quick glimpse, of the feminine world. The other keywords listed for positive keyness that are more expected, such as *http*, *comments* and *www*, are all related to the digital environment. To be more specific, it is highly controversial whether *http* and *www* should be included or not in the keyword list, as they are acronyms (i.e. *hyper text transfer protocol* and *world wide web*) pertaining to the world of information technology.

In Figure 3.4, the list of keywords is listed for negative keyness using the whole ICE GB as a reference corpus.

The first two items are clearly connected to the written nature of LJC. Although the latter shows *some* spoken-like features that have been associated with the digital discourse mentioned in Chapter 1, interjections, such as *uh* or *uhm* are not particularly surprising, as they are unlikely to be found in blog written entries, however informal and conversational they might be. The same may be said about *yeah*, *yes*, *mm* and *m*, that are highly unrepresented in LJC. Enclitics like *'s* and *'ve*, conversely, appear so prominently as they have been transcribed differently in the two corpora.

What difference is displayed from the analyses of the two different ICE GB subcorpora, namely ICE_spoken and ICE_written? A study into keywords is likely to allow some considerations about presumed features of oral or written discourse in the LJ-based blogging communities. It cannot, in fact, be overstated that these observations are meant to represent a foray into one specific blogging environment, given the polymorphic nature of blogging practices that we postulated at the beginning of this chapter. Such polymorphic nature, in effect, is evident from both a diachronic and synchronic perspective; in other words, blogs *do change* in time and in space, as any text does.

As argued in Chapter 1, a lexical analysis should be based not only on isolated words, but also on how words combine and work together.

File   Edit   View   Compute   Settings   Windows   Help

N	Key word	Freq.	%	RC. Freq.	RC. %	Keyness	P	Lemmas	Set
1	UH	19		4,186	0.78	-2,400.(	0.0000(		
2	UHM	4		2,925	0.55	-1,759.(	0.0000(		
3	S	662	0.34	5,716	1.07	-1,039.(	0.0000(		
4	OF	3,623	1.87	15,334	2.86	-591.90	0.0000(		
5	VE	28	0.01	1,061	0.20	-469.45	0.0000(		
6	IS	1,367	0.71	6,862	1.28	-465.01	0.0000(		
7	THE	8,491	4.38	30,172	5.63	-459.38	0.0000(		
8	YEAH	86	0.04	1,418	0.26	-444.96	0.0000(		
9	YES	98	0.05	1,467	0.27	-433.44	0.0000(		
10	WHICH	247	0.13	2,246	0.42	-432.01	0.0000(		
11	THAT	1,964	1.01	8,705	1.62	-397.74	0.0000(		
12	MM	3		486	0.09	-271.55	0.0000(		
13	M	90	0.05	992	0.19	-231.42	0.0000(		

Figure 3.4   Keywords sorted by negative keyness

Keywords are isolated words that appear with unusual positive or negative frequency in comparison to a reference corpus, whereas lexical bundles and word collocations are combinations of words that occur more or less frequently in comparison to lexical bundles in a reference corpus. The way in which words combine can shed light on some issues, for example testing the validity of some assumptions. The high incidence of personal pronouns such as *he*, *him* and *his*, may be interpreted as a specific feature of both fanfiction entries and also of more general fandom-related entries, which consist of descriptions and characterizations of male characters. As a case in point, the first positive keyword with general ICE as a reference corpus, *his*, has been selected for further analysis. Table 3.12 shows three-word clusters sorted by frequency.

Table 3.12   Three-word clusters sorted by frequency

N	Cluster	Freq.	Length	Related
1	SHOOK HIS HEAD	37	3	HE SHOOK HIS HEAD (8), PAPA SHOOK HIS HEAD (6), SHOOK HIS HEAD IN (6)
2	MADE HIS WAY	26	3	AND MADE HIS WAY (8), HE MADE HIS WAY (6)
3	OUT OF HIS	26	3	
4	HIS HEAD ON	16	3	HIS HEAD ON HIS (10), HIS HEAD ON HIS BACK (10), THREW HIS HEAD ON (10),
5	HIS MIND HE	16	3	INTO HIS MIND HE (6)
6	HIS EYES AND	16	3	OPENED HIS EYES AND (6)
7	ON HIS FACE	16	3	
8	HIS FEET AND	15	3	ON HIS FEET AND (8), TO HIS FEET AND (7), ROSE TO HIS FEET AND (6)
9	IN HIS HAND	15	3	
10	HIS HEAD AND	13	3	
11	TO HIS ROOM	13	3	
12	HIS HEAD IN	12	3	SHOOK HIS HEAD IN (6)
13	IN HIS MIND	12	3	
14	THREW HIS HEAD	12	3	THREW HIS HEAD ON (10), THREW HIS HEAD ON HIS (8)
15	ONE OF HIS	11	3	
16	BACK TO HIS	11	3	
17	HIS HANDS IN	11	3	HIS HANDS IN FRONT OF (5), HIS HANDS IN FRONT (5)

In Table 3.12, a sample basic cluster analysis for the top keyword with positive keyness shows patterns of recurrent co-occurrence in the corpus. Even a cursory look at Table 3.12 shows two blocks of usage, one is for prepositional phrases and the second is for noun phrases, typically associated with body parts, such as *head, hand, face* or *eyes,* or mental activities or behaviour, such as *mind* or *smile.*

This seems compatible with the possible overuse of descriptions and characterizations in blog entries in fanfiction, that features long and male-dominated narratives, and also for fandom blogging, which is characterized by a high number of entries devoted to TV and movie (male) stars.

Keyness is a textual notion and keywords can be "globally spread or locally concentrated in bursts" (Scott and Tribble 2006: 66). By the same token, linkages between keywords as collocational neighbours are a textual concern that is interesting to explore. Items either side of any given keyword, considered as a node, link to other keywords which not only share textual keyness, but also local proximity (Scott and Tribble 2006: 67). In Table 3.13, links to the top keywords rated for positive keyness are listed. Links are co-occurrences of keywords within a default collocational span (Scott 2012: 192), which in this case is five words to left and right of the node. Table 3.13 shows the first top ten keywords sorted by frequency

Table 3.13   Keyword links in LJC

Keyword	Links	Hits	10 most-linked keywords
said	158	714	eyebrow, women, room, beer, cloak, never, clothes, book, murmured, walked
new	91	237	inside, against, sci, story, gets, prometheus, glee, game, started, romance, drama
man	83	277	wants, mother, stone, gets, drama, make, watch, took, room, new
little	80	265	night, kid, maybe, kids, make, throat, everyone, dean, surprise, lips
comments	78	309	suck, japanese, thanks, series, authors, beautiful, secrets, hey, everyone
want	74	327	everyone, interrupted, pretty, child, boy, color, drink, even, mommy, sure
eyes	70	218	here, caught, behind, still, ran, chest, cast, me, seeing, who
head	70	226	like, turned, she, over, it's, still, off, bent, when, daddy
make	57	260	he, wanted, little, tell, women, man, anyone, something, after, bad
asked	55	184	it, while, you, up, that, my, looked, parents, after, get

followed by the top ten lexical *linked types*, by the total hits of the keyword (the number of instances of the keyword type) and each individual linked keyword. Looking at the first row, *said* is found 714 times in the corpus and within the collocational span five left to five right, 158 other keywords have been found. The item *said* features 158 proximally located links (i.e., 158 different keyword types) and in Table 3.13 only the ten most frequently linked are reported. As Scott and Tribble suggest (2006), the number of hits has little to do with keyness, because "a word can be key even if it occurs only a few times" (2006: 67). The most frequent words are, more often than not, function and not lexical words. Furthermore, linkedness is related to frequency in general, and the more occurrences of each keywords are in the text, the more likely it becomes that other keywords will be found as linked. For example, *want* occurs 327, so it will be very likely that this item will contract more links than the item *drama*, which occurs 46 times. In effect *want* contracts 74 links, whereas *drama* contracts only 26. "Keyword linkages are not like ordinary collocational linkages since they require both node and collocate to be 'key'" (Scott and Tribble 2006: 68). This further analysis may illustrate which items are *most frequently* linked to keywords, thus providing another layer of exploration into blogging textuality.

3.4.2 Lexical Bundles in LJC

Clusters are words that tend to co-occur. They are also called lexical chains or multi-word units and are a useful tool for lexical studies in general (Stubbs 2002). Multi-word units have been studied as *lexical phrases, routines, formulas, fixed* or *frozen expressions* and *pre-fabricated units*. Different approaches tend to interpret multi-word units according to various criteria, for example considering them as idioms (e.g., *the heart of the matter*) or as non-idiomatic but perceptually salient expressions (e.g., *you're never going to believe this*).

The approach adopted in this book has been drawn from Biber et al. (1999), who call these word sequences *lexical bundles*. This definition was used in the *Longman Grammar of Spoken and Written English* (Biber et. al 1999), which compared the most common occurrences in conversation and academic prose. Biber and Barbieri define them simply (2007: 264) "as the most frequently occurring sequences of words, (e.g., *I don't know if, I just wanted to*)." In their view, "lexical bundles are usually *not structurally complete and not idiomatic in meaning*, but they serve important discourse functions in both spoken and written texts" (Biber and Barbieri 2007: 264, emphasis mine).

Biber, Conrad and Cortés (2004) developed a functional framework for the heuristic description of lexical bundles in discourse, distinguishing three subtypes: *stance expressions, discourse organizers* and *referential expressions*. Stance bundles express attitudes, judgments or assessment of certainty that frame some other proposition, discourse organizers establish a relationship between what comes before and what comes after, and referential bundles identify entities or parts of entities.

Following Biber and Barbieri (2007), lexical bundles have been identified in this study using a frequency-driven approach. In other words, they are simply the most frequently occurring sequences of words in LJC that includes only one kind of text, namely written blog entries. The cut-off point in this study is of 30 occurrences per 100,000 running words. Three-word sequences have been analysed, partly due to the small size of the corpus. The overall LJC has been reduced in this analysis, as multiple entries by a single author were excluded to avoid the occurrence of idiosyncratic usage by the same blogger.

As said, lexical bundles are different from other idioms and fixed expressions, because they are not idiomatic in meaning and not perceptually salient. Furthermore, they do not usually form a complete structural unit. For example, Biber et al. (1999) found that only 15 percent of the lexical bundles in conversation can be considered as complete clauses or even phrases and less than 5 percent of lexical bundles in academic prose form complete structural units. Instead "most lexical bundles bridge two structural units: they begin at a clause or phrase boundary, but the last words of the bundle are the beginning elements of a second structural unit" (Biber and Barbieri 2007: 270). According to the authors, most lexical bundles in speech bridge two clauses (e.g., *I want to know, well that's why I*), whereas lexical bundles in writing usually bridge two phrases (e.g., *in the case of, the base of the*).

Despite the fact that bundles are neither idiomatic nor structurally complete, they have several functions in discourse, for example framing new information, either providing clues as to what the speaker/writer stance is or organizing the textual structure of the text.

To analyse the main three-item bundles in LJC, an index was created using the Wordlist function of Wordsmith 6. An index was generated and then clusters computed, setting 3 as the cluster size (contractions are counted as one unit, i.e., *don't*), 30 as minimum frequency and 50,000 as maximum percentage frequency, stopping at sentence breaks, following the default settings of Wordsmith. In Wordsmith, lexical bundles are called clusters, as Scott terms them (Scott 2012: 337).

As a starting point, we need to make clear that three-word lexical bundles are very common in language, both in spoken and written texts. Biber et al. (1999) claim "three-word bundles occur over 80,000 times per million words in conversation; over 60,000 times per million words in academic prose" (1999: 994). Four-word lexical bundles were, for this reason, selected for analysis in Biber et al. (1999), but the present work has a much more restricted agenda and is based on a much smaller and specialized corpus. However, changing these initial settings significantly alters the results, which may be nonetheless symptomatic of some major trends in blogEng in the *LiveJournal* community.

Biber, Conrad and Cortés' taxonomy (2004), which categorizes three major functional groups and other more delicate sub-categories for bundles, was used to interpret LJC dataset. To this purpose, concordance listings of individual bundles were generated for an in-context analysis.

3.4.2.1 Stance Bundles

Stance bundles express epistemic evaluations and attitudinal/modality meanings and include: epistemic, desire, obligation, intention/prediction and ability. Some examples taken from the concordance listings of the most frequently occurring bundles found in LJC can be seen as follows:

Epistemic Bundles:

> *I don't know* what it feels like
> *I don't know* how to import .mov or .flv
> . . . describes accurately *the fact that* if we truly love something. . .

Desire Bundles:

> *I don't want* to do it again!
> . . .*I just want* to talk about something first
> *I want to* ask you something

Obligation Bundles:

> It *doesn't have to* be a professional publication
> . . .you *don't have to* go anywhere during the day

Intention/Prediction Bundles:

> . . .it's definitely *going to be* a fight
> That was when I knew this was *going to be* a terrible customer

Ability Bundles:

> . . .and I don't think i'd *be able to* get anyone to move. . .
> . . .annual feeling of dread just to *be able to* make my vacation plans.

3.4.2.2 Discourse Bundles

Discourse bundles serve the purpose of framing overall discourse structure and include topic introduction, topic elaboration/clarification and referential identification/focus. Here are some examples taken from the concordance listings of the most frequently occurring bundles in LJC:

Topic Introduction Bundle:

> Given *what I found* when I searched. . .

Topic Elaboration/Clarification Bundle:

> It *has to do* with the previous comment. . .

Referential Identification Focus Bundle:

> ONLY *for those participants* that posted last week

3.4.2.3   *Referential Bundles*

Finally, referential bundles identify an entity or select some specific feature of that entity which is deemed as particularly important and include imprecision bundles, bundles specifying attributes and time/place/text-deixis bundles. Some examples from LJC are reported here:

Imprecision Bundles:

> . . . or they have *some sort of* discomfort. . .
> . . . or maybe *something like that*.

Bundles Specifying Attributes:

> . . .at least *a couple of* hours. . .
> I'm interested in something *a little more* tangible.
> . . .right *in the side* of his big, annoying head. . .

Time/Place/Text-deixis Bundles:

> . . .*on the other* side of the road.
> You could only find out *at the end* of the summer. . .

What is the frequency of these kinds of bundles in LJC? Limiting our analysis to the top 15 bundles with cut-off point of minimum frequency 30, the following bundles were identified:

1. *One of the*: bundle specifying attributes
2. *In front of*: place bundle
3. *Out of the*: place bundle
4. *A lot of*: imprecision bundle
5. *Be able to*: ability bundle
6. *Do you want*: desire bundle
7. *I don't know*: epistemic bundle
8. *I want to*: desire bundle
9. *You want to*: desire bundle
10. *I don't think*: epistemic bundle
11. *I wanted to*: desire bundle

12. *The end of*: bundle specifying attributes
13. *Going to be*: prediction bundle
14. *Part of the*: bundle specifying attributes
15. *The rest of*: bundle specifying attributes

Given the relatively low number of three-item bundles with minimum frequency of 30 occurrences and of each specific instance of bundle (the top bundle appeared 80 times), a concordance listing was carried out for all of them to allow for a more fine-grained analysis.

As a consequence, only the bundles which were functionally fitting the above mentioned categories were included, excluding occurrences that were not pertaining to the category, such as, for example, the occurrences of *at the end of the day*, that is not considered here as a lexical bundle, but as a fixed expression, or all cases where *going to be* did not signal prediction. Another example is *one of the*, which only in two occurrences was followed by *reason: one of the reasons* was best described in those two occurrences as an identification/focus bundle.

Consistent with this methodological framing, those two occurrences were excluded, re-computing bundle frequencies. The usefulness of concordance listings was apparent in all those instances, as they helped clarify the meaning in context. All the bundles thus analysed, were re-arranged in order of frequency, adjusting the Wordsmith results in order to include only those bundles that fitted the specified functional category. Grouping them further into the three major categories as described above, it appears that LJC1 includes:

1. 8 stance bundles (1 ability, 4 desire, 1 prediction, 2 epistemic);
2. 0 discourse bundles; and
3. 7 referential bundles (2 place, 1 imprecision, 4 specifying attributes).

With regard to the number of actual occurrences within the corpus for each of these functional categories, stance bundles amount to 256 occurrences per eight kinds of bundles, whereas referential bundles amount to 309 occurrences per seven kinds of bundles. The top four bundles are referential and this explains the fact that despite their lower number, if compared to stance bundles, their incidence is higher.

These findings may be useful in further exploring the *spoken/written* nature of blogEng. Again following Biber et al. (1999: 991), we learn that "in conversation, for example, a large number of lexical bundles are constructed from a pronominal subject followed by a verb phrase plus the start of a complement clause, such as *I don't know why* and *I thought that was.*" In academic prose, conversely, lexical bundles are more frequently parts of noun and prepositional phrases, such as *the nature of the* and *as a result of*. They also contend that most bundles in conversation include the beginning of the main clause, followed by the beginning of an embedded complement clause (Biber et al. 1999: 991).

In Halliday's terms, many lexical bundles in conversation include the beginning of a mental clause plus the beginning of a projection clause. An
important generalization on the matter thus concerns a major difference
between lexical bundles found in conversation and in prose: most bundles
in conversation are building blocks for *verbal and structural units*, whereas
bundles in prose are building blocks for *extended noun phrase and prepositional phrases* (Biber et al. 1999). "Epistemic bundles are relatively common
in spoken registers" (Biber and Barbieri 2007: 274).

With regard to formal aspects, a comparison between our findings and
Biber et al. (1999) show that some bundles are typically more likely to occur
in conversation, whereas other less frequent are more typical of academic
prose. In the following section, bundles from our datasets that may be ascribed to either conversation or prose will be categorized.

3.4.3.1   *Conversation*

Although most lexical bundles cannot be analysed as complete structural
units, they have been grouped also into categories that take into account
their structural correlates. In conversation, the most frequent category is
represented by a personal pronoun + lexical verb phrase (+ complement-
clause fragment). This structure is also very common in our dataset, as the
bundles *I don't know*, *I want to*, *you want to*, *I don't think* and *I wanted to*
attest. The bundles in this category display a first or second person pronoun
as subject plus a stative main verb. Many of them are used to begin a clause
and end with the beginning of a complement clause: either a *that*-clause
(that may be also omitted), a *to*-clause or a *wh*-clause. In this respect, the
bundle is increased to include four, five or even, very rarely, six items, as in:
I don't know who (to credit), *I don't know why you'(re here or anything)*, *I
don't know what it's like*. As said before, all bundles of this category express
stance in LJC, as they convey meanings such as personal opinions, feelings,
desires and thoughts. The first person pronoun is the most frequent, attest-
ing bloggers' strong commitment towards their own texts (i.e., entries). It is
interesting to note that there are also two negative epistemic bundles, namely
I don't know and *I don't think* that may be interpreted along the same lines.

There is only one instance of a second person pronoun (i.e., *you want to*),
however, consistent with Biber et al.'s findings (1999), it is usually part of
larger interrogative or conditional clauses, as in:

> If *you want to* add text to your gifs, didn't *you want to* go back?
> If *you want to* ignore the discussion, why do *you want to* know?

This is also attested in another more frequent bundle, *do you want*, that is
part of *yes-no questions* fragments, which begin with either an auxiliary or
a modal verb, followed by a subject pronoun. This bundle is the initial part
of a *yes/no question*.

Another bundle that is listed among the conversational categories is the modal expression *going to be*, as in:

I'm *going to be* looking after the prince. . .
. . .it's definitely *going to be* a fight. . .

Obviously, modal and semi-modal expressions in general, and this specific bundle in particular, are special, because they are rather fixed and relatively idiomatic. Their meaning comes consequently as a "block" and cannot be understood by analysing its separate components.

The bundle with *to*-clause fragment is represented by *be able to*, where *able* is followed by a *to*-clause, as in:

I would *be able to* pay interest. . .
. . .we'll *be able to* see the lake. . .
I'm envious that Daddy always seem to *be able to* make Mommy so happy.

Among the noun phrase expressions we have the beginning of a noun phrase plus the preposition *of*, as in *the end of* and *the rest of*. The former usually specifies time or place, as in:

. . .you could only find out at *the end of* the summer. . .
. . .it might only be for a second at *the end of* the season. . .
Justin reached *the end of* the rope. . .
. . .try shaking it on blotting paper at *the end of* the line. . .

It is worth noticing that this bundle is, in the overwhelming majority of cases, part of a four-word or in few cases (i.e., 9) five-word bundle, as in *at the end of the*. . . .
Prepositional phrase expressions, such as *out of the* and *in front of*, are also very common in conversation and begin with a preposition plus a noun phrase containing *of*. They are used in most cases to indicate place, as in:

. . .he stepped *out of the* building. . .
. . .he rushed *out of the* room. . .
. . .we can force you *out of the* workplace. . .
. . .when he was *in front of* the building. . .
. . .a new neighbor move *in front of* us. . .
. . .the boy *in front of* him suddenly froze. . .

3.4.3.2 Academic Prose

Only two bundles can be associated with academic prose, that is, *one of the* and *part of the*. The former is a noun phrase with *of*-phrase fragment, as in:

I think she's *one of the* most complex characters. . .
. . . a Law student at *one of the* most famous Universities. . .
. . .*one of the* schools that offer breakfast. . .
. . .and *one of the* most amazing things. . .

Part of the is also a noun phrase with *of*-phrase segment, as in:

. . .will always be *part of the* "Top Model" family. . .
. . .the *part of the* male brain that worries. . .

These bundles cover a wide variety of functions, but our dataset exhibits most occurrences of bundles that give descriptions of various kinds (e.g., physical, metaphorical, temporal, etc.). Also in these two bundles, most instances display four-word bundles, such as in *one of the most, one of the main,* and *the part of the.*

Such prepositional bundles are very common in both spoken and written texts and their role is to specify places, time, size, amount and characterization in general. Accordingly, bundles, such as *the end of, the rest of, one of the* and *part of the*, have been classified in separate categories and into separate slots, with the former two fitting into conversational bundles and the latter into academic prose bundles. However, these are overlapping categories in many cases and cannot be considered as more or less associated with either speech or formal writing.

One of the conclusions we can draw from this analysis on lexical bundles is that stance bundles are largely present in LJC dataset. Noun and prepositional phrase expressions are also very common, but it is interesting to note the high incidence of lexical bundles expressing a blogger's stance, attitude and position. The small dataset does not allow for generalizations, but a comparison between the findings obtained through a keyword analysis and a lexical bundle analysis are fully compatible. What appears is that blogEng shares some features with more "spoken-like" text genres and some features with more "written-like" genres. It lies somewhat in-between the *continuum*. Variation is, thus, better described in terms of its positioning along some of the dimensions identified by Biber (1988). A lexical analysis complements such findings in that it shows that many bundles (i.e., words that co-occur together) are found more frequently in conversation, whereas those bundles that may be associated with prose are rarer and can also be found in conversation as they simply signal, say, attributes of places and entities and, as such, are very general chunks of meaning in a broad range of communicative situations. What is then the relationship between these different features and their use within the blogEng domain?

The distribution of such features is quite varied, but some of them are consistent in many different datasets: for example, the frequency of stance bundles is noteworthy and points to the highly personal nature of blogEng entries, which consistently present the first person pronoun as a subject who thinks, believes, and expresses opinions in the digital community.

3.5   INTEGRATION OR COMPARTMENTALIZATION? SPEECH PARTICIPATION VS. WRITING SECRETS

Blogs have been defined as a polymorphic genre in this chapter, for a variety of reasons. First of all, the blogosphere has been constantly changing since its appearance in the online world. There are many reasons for thinking that blogging practices are, today, approaching the sphere of the digital world where social networking communities live and thrive. This is why, in this chapter, a corpus of entries from a social platform, *LiveJournal*, was collected to see transformations and blogging attitudes within a representative environment of many of the transitions that we are witnessing in the wake of the 2010s.

A study on blogEng has been grounded on the consideration that much previous research literature has focused on issues such as credibility and relationships with mainstream media. Today, however, some communities of practices within the blogosphere are becoming more and more embedded within social networking activities, such as commenting, sharing hobbies and interests, writing on everyday activities, but also talking about common passions, such as fandom and fanfiction. Special interest has been placed on such blogging practices, also considering that the language used in fandom and fanfiction, that is typical of fan communities, is still relatively unknown. Fanfiction encapsulates a huge number of narratives that could have justified the preliminary assumption that such narratives were firmly grounded on a purely written tradition. However, this assumption has been disconfirmed by crisscrossed studies on different datasets that conversely show a significant prominence of personal and highly context-dependent narratives. Such narratives prove inaccessible, or very partially accessible, to anyone who does not share the digital context of situation. Very high reliance on context-dependent features may indicate the preference for content production that is inaccessible for non-acolytes.

Some well-established ideas, such as the much-heralded concept of the integrability and integration of semiotic resources on Web 2.0 digital environments, have been challenged, as preliminary findings show that such integration is not as tight as it was supposed to be. Findings, for example, show that most entries relying on the resource of writing consistently tend to employ few other resources. The same can be said about those entries mainly relying on other resources, for example visuals, that tend to manage without writing, that is, in these cases, employed as a mere subservient of salient visual resources (e.g., use of captions for pictures or drawings). Resource integration thus seems to be less prominent than could be assumed, opening further research questions, such as the applicability of this theoretical claim to other blogging platforms, for example changing dependent and independent variables.

Despite the relatively small size of the corpus, findings shed light on language mode variation across entries. The corpus has been selected to include a range of entries from different popular blogs (or journals). A keyword analysis has shown mainly three different aspects: (1) keywords concern digitally-related contents (e.g., *comments*, *posts*) and personal nouns; (2) if compared to a general corpus of British English, items that may be related to forms of narrative characterization appear as positive keywords (e.g., *his*, *him*, *her*); and (3) if compared to the reference corpus spoken and written subcomponents, entries may be defined as *in-between* texts, as they exclude both extreme spoken-like features, such as interjections, but also extreme written-like features, such as the definite article *the* and *of*-clauses that are specifically found in fully-sustained prose. Some items have also been analysed partially following an MF/MD analysis, to test whether they have positive or negative loadings along the Dimensions that have been previously devised to explore spoken-written variation (Biber 1988).

The corpus-based approach has also been applied to analyse the top fifteen three-word lexical bundles in a selected dataset with the purpose of investigating them from both a functional and formal point of view, within the general agenda of placing such bundles within the spoken and written *continuum*. Consistent with previous keyword analyses, the most frequent bundles are *stance bundles*, expressing bloggers' attitudes, point of views, and positions. Discourse bundles are less frequent, whereas referential bundles appear more consistently in the corpus. However, it must be said that three-word referential bundles are very common in both spoken and written genres and, as such, their placement within either of them is neither definitive nor univocal.

All these observations do not provide ways of interpreting in stable or permanent terms the blogging experience. The present work has been mainly focused on specific kinds of blogs, particularly those of a personal nature. However, in the wake of the 2010s, the blogosphere is undergoing another profound transformation witnessed by the transition towards forms of communal and participatory culture, also instantiated in micro-blogging that reflects questions discussed in Chapter 1.

Nowadays, blogs tend to become conglomerates instead of single interconnected and linearly distributed points of access to hyperselves and hypermedia (Thibault 2012). The chapter's broad agenda has been to capture blogging practices in their polymorphic nature, hence not stable for definitive characterization, as is ultimately the case with any web-based genre, if not *any* text genre. This chapter raises more questions than it answers, but some findings may prove useful for further analyses. The common ground of all these observations and reflections is the notion of polymorphism that has proved invaluable to providing several points of access to a world that is less open and democratic than it might seem at first sight.

Referring back to the questions raised at the beginning of this chapter, the implicit interactivity of blogs is countered by the individual contribution that is shaped across different and diverging lines: hierarchical control of contents by bloggers is grounded on notions of authorship and provenance that are ultimately connected to blogs positioning within corporate mainstream media. Such positioning—be they ideological or profit-based—are not always easy to detect, but the blogger's stance is a key feature to understanding how such volatile pieces of writing (or speech, in the case of videoblogs) are ingrained into individual and collective social practices. The relationship with mainstream media has been amply documented in research literature, but is now becoming a misguided question, as more and more blogs today are encapsulated in corporate mainstream journalism. Furthermore, debates about blog formality or informality have been losing ground, as they are more than ever fragmenting into a galaxy of different constellations.

The evanescence of individual writing of individual bloggers, grappling with their lives and building up personal narratives, mixes with the permanence of their own social practices within the digital world. The credibility of their stories, be they personal or political or both, is instantiated in the credibility of the medium that mainly lies in writing, revealing its secrets only to those who already know them.

Appendix 3.1

ICE: Great Britain Corpus Structure

The sampling structure of the corpus is shown here.

Spoken Texts (300)	Dialogues (180)	**Private (100)**	face-to-face conversations (90) phone calls (10)
		Public (80)	classroom lessons (20) broadcast discussions (20) broadcast interviews (10) parliamentary debates (10) legal cross-examinations (10) business transactions (10)
	Monologues (100)	**Unscripted (70)**	spontaneous commentaries (20) unscripted speeches (30) demonstrations (10) legal presentations (10)
		Scripted (30)	broadcast talks (20) non-broadcast speeches (10)
	Mixed (20)		broadcast news (20)

Written Texts (200)	Non-printed (50)	Non-professional writing (20)	untimed student essays (10)
			student examination scripts (10)
		Correspondence (30)	social letters (15)
			business letters (15)
	Printed (150)	Academic writing (40)	humanities (10)
			social sciences (10)
			natural sciences (10)
			technology (10)
		Non-academic writing (40)	humanities (10)
			social sciences (10)
			natural sciences (10)
			technology (10)
		Reportage (20)	press news reports (20)
		Instructional writing (20)	administrative/regulatory (10)
			skills/hobbies (10)
		Persuasive writing (10)	press editorials (10)
		Creative writing (20)	novels/stories (20)

Taken from http://www.ucl.ac.uk/english-usage/projects/ice-gb/design.htm. Last accessed June 17, 2012.

4 Interacting on a Global Scale
Speech and Writing in YouTube Comments

4.1 FORMS OF TRANSITIONS

This chapter discusses textual transformations and transitions in the participative culture of web-based media sharing communities and in their textual
practices, such as commenting activities and video responses. Participative
culture in media sharing communities is connected with language mode
variation in conflicting dynamics, for example: interactivity *versus* isolation,
authorship *versus* anonymity, verbal exchanges *versus* video conversations,
content evanescence *versus* content stability, performance *versus* informa-
tivity, repetition *versus* creativity, self-representation *versus* self-disguise.

The strategies of personal expression and performance within the textual web galaxy are continuously subject to changes. Thus, they do not
allow for stable theories or for the elaboration of models of analysis belonging to one specific field of studies. Exhibiting oneself through complex
socio-semiotic strategies is within easy reach for web users, who have the
unprecedented opportunity to harness linguistic, visual and multimodal
strategies for self-representation. However, the creation of virtual identities,
avatars and nicknaming hinges on the construction of *alternatives to the self*
through multiple semiotic and discursive stances, which will be examined
through multimodal theories and corpus-based lexical studies. The purpose
is to analyse discourse practices used for the *mise-en-scène* of the self, which
fluctuates between attempts at authentic self-representation, performance
and fiction. As in the previous chapter, a corpus-based lexical approach
will be applied to study verbal comments, while bearing in mind that these
verbal comments are embedded in multimodal environments and that it is
within the multimodal domain that the written/spoken interplay develops
its epistemological and semiotic specificity.

In the previous chapter, the study on blogs was focused on one community gathering personal journals, *LiveJournal*. As discussed, blogs have been
entering the phase where separate and distinct blogs formed a constellation
or *blogosphere*. Blogs have since been incorporated into wider communities
of interaction. Furthermore, defining blogs as belonging to a genre is problematic, as they are embedded in social networking communities. Research

literature has focused on blogging in relationship with, and in comparison to, mainstream media, and in particular mainstream journalism. However, current political blogging, originally equated with independence and freedom in writing, is now adopting corporate and profit-based status to survive or, in other words, *to be read*. Freedom of expression proved difficult to survive in independent blogging, so other alternative forms have been developed.

Furthermore, the common view of the high integrability of semiotic resources has been challenged in the previous chapter, as they appeared as less integrated than expected in *LiveJournal*. A focus on the interplay of the written and spoken language mode has also shown that web-based micro-texts, such as blog entries, are hybrid creatures, in that they exhibit both spoken-like and written-like features of discourse. Classifications thus become unstable, even illusory, in that such texts present rather unpredictable configurations in meaning-making events, and well-established notions, such as those of web genres, become blurred and difficult to pin down.

Micro-textuality is a recurring definition in this book. Micro-texts constellate the web and inform the interplay between speech and writing. They are not only characterized by their brevity (as in the case of chat messages, blog entries, status updates in social networking websites or comments in media sharing communities), but also by the *definitional* alternation between spoken-like and written-like features, as the previous chapters have evidenced.

The extent and range of the occurrence of uploading video on self-representation will be studied using a case study from YouTube, a video sharing community that produces more than 10,000 broadcasting hours per day, corresponding to 24 hours a day broadcasting of 400 TV channels (Wesch 2009a). Snickars and Vonderau (2009) argued that YouTube was growing at the amazing rate of 75 percent per week in 2005 and that by mid-2006 the site had 13 million unique visitors every day, watching more than 100 million videos.

The opportunity to publish original and self-produced texts on the web amplifies the overexposure of personal identities, strengthened by the relationship between videos and related comments, which will be analysed in the dialogic dynamic of *self-representation* and *authenticity* that are two facets of spoken/written variation. As highlighted with reference to debates on the interplay between orality and literacy discussed in Chapter 1, mode use and variation are socially and culturally constrained. The material constraints of the medium are also relevant, as they have an impact on how a text is built and negotiated across different texts and cultures.

In this chapter, the methods used in Chapter 3 have also been adopted. A corpus of YouTube comments was built and explored using tools of corpus linguistics to investigate lexical patterns symptomatic of variation between speech and writing. The multimodal embedding of such micro-texts emphasizes the hybrid and polymorphic nature of such texts, in a similar vein to that found in Chapter 3. Far from being two separate and parallel ways of tackling the issue of web-based textuality, multimodality and corpus linguistics will be used as complementary tools and consistent theories to provide a coherent framework of analysis for highly unstable and richly semiotic texts.

4.1.1 YouTube on the Global Communicative Mediascape

YouTube is a video sharing social network where users upload, share and view videos. It was created in 2005 by three friends, Steve Chen, Chad Hurley and Jared Karim, former PayPal employees, who claimed that the idea of creating a system of video sharing on the web came to their minds after a party apparently unattended by one of the three, Karim, who denies it ever occurred. It is Karim who is credited for the early idea of an online video sharing community, even though he subsequently claimed that the realization of YouTube required "the equal efforts of all three of us" (Cloud 2006).

YouTube's functionalities include the view of commercial and promotional videos, such as music videos, trailers, clips and almost countless user-generated videos. The slogan "Broadcast yourself" gives a clue about the phenomenon which, in the last few years, has been generating a huge mediated and participative culture, based on sharing texts whose extremely complex socio-semiotic nature is unprecedented within the global mediascape (G. Jones and Schieffelin 2009).

Studies on these texts and socio-semiotic affordances of video sharing communities, of which YouTube is probably the best-known example, are still in their infancy. They can be studied from different standpoints and under the lens of many disciplines, such as digital anthropology, linguistics, socio-semiotic and multimodal studies and, more generally, social sciences. The radical transformation of the once-only viewer into producer, screenwriter, actor and broadcaster of their own video contents is an aspect of the more general occurrence of mass media digitalization. This aspect also entails the subversion of the linear and unilateral order of broadcasting *for* the audience and a dramatic reduction of video production by traditional entities, such as the movie industry, public and private institutions, agencies and companies. However, the most significant amount of video production is currently *user-generated*.

Users move within the blurring borders of what may be defined more generally as a *mise-en-scène* of the self. The latter does not have direct commercial purposes, although in some cases the impact of the video is so huge that produces an indirect commercial spin-off, as will be illustrated hereafter. The traditional linearity of the production/reception system allowed two basic reactions in the viewer: (1) *deliberate consumption*, that is, buying the mediated product (for example going to the cinema or buying a CD); or (2) *offer rejection*. Both options exclude the possibility of elaboration of a personal creative space to develop a personal identity and sense of intersubjectivity. They both avoided the user-generated creative breaking of the production/consumption chain. These practices have become an episte-mological possibility, encouraging a revolutionary twist within neoliberal capitalism. It is doubtless controversial whether remixing the old into new elements and the consequent perpetual recycling of contents is valuable or not (Lanier 2010). Hot debates are also concerned with possible risks of flattening identities and paving the way to aggression and lack of commit-ment, due to easy digital camouflage and masking (Lanier 2006). However various considerations may end in approving or disapproving remarks, web culture is contributing to democratic circulation of contents, adding to the proliferation of web texts and genres. The latter need tools of analysis able to probe into text forms, politics, aesthetics, ideologies, ethical and theoreti-cal implications.

Traditional linguistics does not provide comprehensive theories and tools for such an undertaking, whereas other approaches encompass the conscious harnessing of other socio-semiotic resources, thus including the study of audio, visual and multimodal resources (Kress and van Leeuwen 1996, 2001, 2006; Baldry 2000; O'Halloran 2004; O'Halloran and Smith 2011). Socio-semiotic and multimodal approaches are more flexible and likely to account for digital textual complexities, which, by definition, inter-weave multiple semiotic modes and channels, which result in complex and unparalleled outcomes and expressive dynamics. One of the elements here analysed is how the exchange of contents is managed on video sharing com-munities, shedding light on the resulting different text genres and discourse practices which position politics of self-representation across written and spoken variation and multisemiotic representations.

In other words, what has the written/spoken variation to tell us about how participants create their own interactive and creative space for self-representation and self-exposure on media sharing communities (O'Donnell et al. 2008)? The use of different modes is not casual, as this book has shown, and such use cannot be explained only in terms of medium-constrained practicalities or technicalities. Much in the same way as what happened and happens in communities of practices, which place themselves within a specific point in the oral/literate *continuum*, the same may be said about *digital communities of practice*. Systems of representations and expressions of strong personal commitment (see the example of fanfiction in Chapter 3, or evaluative stance in mode-switching in Chapter 2) are affected by mode choice, or, to put the matter in slightly different terms, mode choice instanti-ates and realises different configurations of the self in semantic and semiotic terms. As the definition suggests, media sharing communities are centred on exchange and interaction, as are the other domains of discourse presented in this book, such as videochats and blogs.

Generally speaking, interactivity may be interpreted as an opportunity for self-expression. It is one of the most significant features of digital media, as in social networks, forums, blogs, and so on, and its overuse has generated several attempts to redefine its epistemological boundaries (Seifert, Kim, and Moore 2008). Although the somewhat loose definition that interactivity currently holds, it is nonetheless true that YouTube intensifies the expansion of user-generated contents and consequent manipulation of such texts and materials (Wesch 2009b). The event is new in that most broadcast material is original and produced using very limited, often rudimentary means. Despite the fact that it is difficult to provide reliable data, it has been estimated that in 2008 about 88 percent of the produced material was user-generated (Wesch 2009b).

As reported, YouTube broadcasts about 9,932 hours of video per day, corresponding to the uninterrupted and contemporary broadcasting of 400 channels. User-generated contents alter previous well-established forms of production, organization and distribution of information. Even the concept itself of information has been profoundly modified, whereas digital information and digital resources have had an impact on cultural, political and social domains (Tredinnick 2006, 2008). In digital contexts, the organization of information is not linear any longer and new patterns in the production, organization and distribution of information generate new forms of sociality and new communicative models, thus requiring new or partially new interpretative frameworks. The rapid development of video sharing communities, such as YouTube, is showing that participation is an essential factor in new conceptions of community and socialization. From a *material* point of view, the implied or explicit notion of "social" or "sociability" is extremely controversial when it comes to digital communities. That is to say in communities where a shared context of situation is simply meaningless.

4.1.2 Communities and Socialization: A Digital Divide?

Establishing and maintaining social relationships are basic human needs, also defining the "typical" human way of inhabiting the world. An early view on sociability dates back to the well-known definition by Simmel, who contended that sociability is association for its own sake, devoid of personal or material implications ([1908] 1921, 1949). While in Simmel's definition, sociability is a "pure" form of being together, in classical literature, sociability as a philosophical and anthropological notion is discussed in Tönnies's seminal works, in the communitarianism project and, more recently, in network analysis. Sociability is considered by Tönnies as the *Gemeinschaft* (community) of mind, which is based on a relationship between friends or a level of "intellectual accord" (quot. in Purcell 2006: 140). Significantly, Tönnies believed, "All intimate, private, and exclusive living together, so we discover, is understood as life in *Gemeinschaft* (community). *Gesellschaft* (society) is public life—it is the world itself" (Tönnies 1957: 33).

Among more recent theories that analyse sociability dynamics from specific standpoints, Purcell discusses the work of the communitarian theorist Etzioni (1993), who highlights the importance of events in the creation of social bonds and communities. Since the final goal of communitarianism is moral persuasion, sociability is a key definitional factor.

But what happens when we consider digital communities? Labels clearly need to be redefined in the light of less clear-cut social forms of groupings and related interactional and conversational patterns. Chien Pan et al. claim that "sociability is a social strategy and a technical structure. It can support the common goals and interaction processes in the online communities" (2007: 427).

In contemporary society, sociability is envisioned in a broader sense than that elaborated by Simmel. It comes to include "companionship, *conversation*, and communication of shared interests" (Purcell 2006: 140, emphasis mine). However, we may note in passing that sociability alone is not sufficient to create communities. Furthermore, ideas of association, social bonds, community and social relationships get intertwined with conversation. Network analysis tends to focus on the ways, strategies and consequences of digital communities.

Despite its social quality, the real interaction between users is always *mediated* by a machine, as suggested in Chapter 2. The number of users that form a social community may vary from the bare two to a virtually infinite number . . . *but one machine is needed per every active user*. The number of users is thus indeterminate, whereas the number of machines is determined by the number of users.

A social network may be social in its purpose, intentionality in the aggregation of users to forming a community and socially driven in the development of specific design and production. However, it cannot be overstated that from a material point of view, the interaction is deployed via a machine and no real (i.e., physical) intercourse of any kind is ever established among web-users. This is not to say that it is excluded or impossible in practical terms, but that the epistemological nature of the online social encounter is based on the contemporary *suspension* of a real encounter. In other words, face-to-face and digital encounters exclude one another: if people meet face-to-face, they cannot meet in the digital world *at the same time,* and *vice versa*. This leads to a controversial view as regards the real social impact of online communication and socialization, which are created and made possible by the unavoidable presence of a machine, which, at the same time, embodies and instantiates the impossibility of a synchronous encounter in real life.

YouTube comes half way, in that it includes features which mimic face-to-face encounters (e.g., the possibility of seeing and interacting with the self-represented "other"), but also postulates *in absentia* and asynchronous exchanges, thus negating the *hic-et-nunc* of synchronous virtual interaction. YouTube allows a wide range of settings and models of participation that challenge the notion of static video sharing practices. Participants may be occasional viewers or active video producers or video commentators. Participation is thus embedded in a more comprehensive system of digital

socialization. Language, social practices and the politics of self-representation are, therefore, deeply intertwined with conceptions of privacy (Gal
2002; Gross and Acquisti 2005). What is considered private and what is
considered public and consequently sharable are not as straightforward as
one might imagine. Any uploaded video may be either shared within the
global community or with a definite set of users (e.g., friends or subscribers to the channel from which the video is uploaded). Users can also post
comments referring to each video, discussing and creating an asynchronous
debate within the potential *macro*-community of tubers. User-friendliness in
creating and sharing videos has encouraged the development of an interactive community, whose aim is, among many, a permanent exhibition of the
self, which parallels a permanent verbal and visual commenting activity on
others' "exhibitions."

Videos of self-representation are extremely popular on YouTube and are
embodied textually in different ways, exploiting different textual practices.
The most obvious is vlogging, that is a "video-diary" or "video-confession,"
but self-representation can be also articulated in the display of a talent (in
music, sport, poetry, etc.) and in the digital reproduction of family moments in everyday contexts, akin to home video practices. *Self-representers*
show themselves or talk to a webcam. They do not have a visible audience, but aspire to reach out to far-away people and places in a permanent
dialectic dynamics, where the exhibition of the self creates a looking-glass
effect. Every comment and response video perpetuates textual reverbera-
tion, whose intertextuality is virtually infinite. The intertextuality of video
sharing communities and social networks is constantly in the process of
becoming saturated, because videos, comments, cross-referencing, videos
responding to other videos create a virtuosic postmodern pastiche. However, saturation is always postponed and deferred, calling for a redefinition
of the concept of intertextuality itself, which blurs along previous clear-cut
borders, thus becoming *hyper-intertextuality*.

Wesch (2009a, 2009b) discusses the concept of *hyper-self-awareness* that
is realised through a virtually infinite projection of the self in the global
community, which is neither a public space nor a physically social entity.
The mere sum of private spaces cannot be equated with a global agora,
because digital communities lack the sense of physical aggregation that,
according to several theorists in the field of social sciences, is an inevi-
table ingredient in processes of classic socialization and society (Grusec and
Hastings 2007).

Regarding transnational migration, Rouse has proposed a model to understand the role played by the media in the dynamics within social networks. According to Rouse (1991), the media used by users to meet up and
interact with other users create a "media circuit." A media circuit is not
an autonomous social network, but is a system providing support for the
existence and development of many social networks, encouraging and technologically mediating social interaction among web-users (Lange 2008).

The examination of a media circuit helps understand the functioning of social dynamics within a given digital community. On YouTube, for example, users can set different levels of privacy, allowing the view of one's own videos to pre-set groups of friends or subscribers. Access is thus highly customizable, according to different ideas of privacy. According to Lange (2008), tubers have different expectations as regards which information can be shared and which data are sensible and thus need to be protected. Lange claims that the concept of fractalization in the public and private sphere is a model that can be employed to study the differences in expectations regarding personal identity and sharing of contents on social networks (Feder 1988; Gross and Acquisti 2005). The fractal model is useful to define the different phases and gradations, which classify levels of the public and private spheres. Such models cannot be interpreted as a polarization of opposed elements, but as a *continuum*. Setting options for YouTube channels allows for a very sophisticated customization and management of privacy, so that levels of sharing can be created to satisfy different perceptions and expectations about the extent and range of content sharing.

Self-representation on YouTube is linked to the consideration and conception of privacy. Self-representation does not necessarily mean self-display: numerous examples (e.g., YouTube celebrities, such as *MadV* and *guitar90* in Figures 4.1 and 4.2, or anonymous tubers) prove that disguise and camouflage are widespread practices on media sharing communities. Self-representation through videos cannot be equated with a transparent show of what one is, providing personal details such as names, surnames, location, and so on, or unveiling a physical part of the self, which defines the self univocally, that is, one's own *face*. Authenticity, or the perception of authenticity, can be transmitted through semiotic and multimodal textual practices, which do not entail the unveiling of the self. In other words, what really matters is to create a consistent and reliable version of the self, for example making oneself recognizable through a consistent and coherent disguise or message in all the videos uploaded in one's own channel (e.g., *MadV*, Figure 4.1) or by foregrounding a talent or an ability, leaving personal details in the background (e.g., *guitar90*, Figure 4.2).

The issue of the supposed falsity of the videos is an important one for the inhabitants of digital communities, such as social networks. Those who upload contents on YouTube and present them as *authentic* and *true*, when this is not the case, are stigmatized and contested by the community. The example of *lonelygirl15* is a case in point. She is a vlogger who introduced herself as a sixteen-year-old girl. She attracted thousands of fans for months and uploaded videos telling about her personal life, casually and spontaneously talking about events and emotions from her bedroom through a webcam. The difficulty in ascertaining whether her vlogs were a hoax created an investigation on a global level. In other words, detailed analyses of her videos, which were viewed and commented by tubers from all over the world, tried to test authenticity and find clues that could unveil the fictional

Figure 4.1   *MadV*, YouTube celebrity director and illusionist

Figure 4.2   *guitar90*, supposed nickname of the anonymous Korean guitarist Lim Jeong-hyun

nature of *lonelygirl15*'s videos. After months of hectic exchanges, the truth was revealed: *lonelygirl15* was Jessica Rose, an actress from New Zealand. The creators of *lonelygirl15* posted an explanation on her vlog, claiming that they were experimenting with a new narrative and interactive genre, and that the line between fandom and stardom had been definitely removed (Heffernan 2006). As a consequence, fans abandoned the web-based show, posting comments that showed their contempt for the hoax and describing themselves as *victims*. Early reactions on the part of fans were as follows:

JayHenry:	"Boooo! Weak"
west:	"This is like R2-D2 poking out and saying 'hey look I'm just a midget' at the end of the ceremony during the finale of star wars."
morninglory:	"That sort of made all the fun come crashing down . . . They call it a story, completely avoiding the fact that it's main appeal was the MYSTERY . . . I almost feel like they thought we'd be honored to know, as if we'd feel like a part of the show.' "
tempestarii, the Thelema expert:	"who is interested in a story when all your investment in the characters within is false"
Hyemew, the Dauntless Dean of All LG15 Fans:	"I knew it was fake, but I didn't want THEM to tell me! Investigation seems a little less fun now that all parties involved acknowledge its fake."

The last comment implies that a part of the fun lied precisely in the mystery and, once all parties acknowledged that it was fake, the fun was definitely over. The experimentation was successful as long as the lies were kept alive as if they were real: once that fiction was revealed, the fans lost interest. Such responses show that the audience tolerates being "cheated" by traditional media, whereas people expect to find "reality" in the new media. Cheating is considered an unacceptable slight (Heffernan 2006).

Authenticity and disguise, confession and falsity, exhibition and *mise-en-scène*. All these forms may be intertwined in multimodal texts, following plots whose textuality is made even more complex by the system of comments, which make the *master-text* (i.e., the video) as a mould for further virtually infinite discussions and debates. Such comments are multi-shaped "adjuncts" to the master-text and shed light on the other face of representation, that of the viewers. Comments are texts that discuss other texts; they are *meta-texts*, adapting from Genette, who devised this category to describe relationships among *literary texts* (Genette [1982] 1997).

YouTube viewers are different from the audience of traditional media (e.g., TV and cinema), as the former incorporate *additional affordances*. In other words, they view a multimodal video using interactive strategies, such as the production of *meta-comments* to the master-(video)text. Among other response strategies and practices, we may also mention the *response video* that is a video responding to the master-videotext through remix or remake practices.

The next section will tackle the case of *viral videos* to shed light on the dynamics of production and reception of multimodal texts, also considering issues such as self-representation and representation of others, borders between public and private sphere and ideas of authenticity and interest, as they are linguistically expressed on verbal and audiovisual levels.

4.1.3   Viral Videos, Privacy and Popularity

Most videos are, as said, user-generated, therefore a range of different phenomena have been developing as a consequence of popularity and global spread of some YouTube videos. For example, viral videos and videoblogging can coincide, because a vlog (i.e., a video-diary) could become a viral video. The latter turns out to be very popular and its popularity is uncontrolled and exponential, regardless of the producer's original intention. It is not easy to discern the real reasons for such a "disease." According to the most rated lexical entry in the Urban Dictionary to date, a viral video is a digital clip "which gain[s] mass popularity through Internet Sharing, such as entertainment websites, e-mail messages or suggesting a friend watch it. Heavy.com and YouTube.com are two well-known examples of media sharing websites which contain viral videos."

An analysis of viral videos can establish a relationship between "virality" and practices of self-representation across mode variation. Viral videos become popular on a global scale. An additional feature of these videos is the production of a huge corpus of comments. Viral videos have originated global celebrities, such as in the case of Justin Bieber, who rose to stardom after a video upload of his singing his pop songs. His video "Baby" is the most viewed ever on YouTube, with more than 810 million views by the end of December 2012.

The following case study has been selected as a model for the exploration of relevant aspects discussed so far, such as the politics of authenticity and self-representation, which are linked to different ideas of privacy, here intended as ways of managing private contents in a public space. Representability is in such cases intertwined with privacy, which underlies the *mise-en-scène* and discourse practices employed to highlight this relationship.

4.2   *CHARLIE BIT MY FINGER—AGAIN!* A CASE STUDY OF A VIRAL VIDEO

The viral video "Charlie bit my finger—again!" features two British children, as represented in Figure 4.3, interacting for fifty-six seconds. The video was shot by their father, who claimed that he wanted to share his children's daily routine with their godparents living in Colorado. This is the most viewed non-commercial video of all time, and one of the most commented and rated on YouTube. At the moment of the last editing of this book, the video had been watched more than 500 million times, featuring almost 800 thousand comments in a range of languages, such as English, Spanish, French, Italian, German, Ukraine, Japanese, Hindi, Urdu, Arab and Tamil, to name but a few.

The video is homemade and represents the exhibition of what is most private, that is two very young children in an everyday context, and was uploaded on May 22, 2007. One-year-old Charlie is sitting on his three-year-old brother Harry's lap, when he bites Harry's finger. The eldest cries, saying: "Charlie bit me . . . and that really hurts!"

The video soon became viral, reaching every corner of the globe and originating a plethora of comments and hypotheses on the creation of the video itself, which some tubers speculate may actually be a hoax. The possibility of it being a hoax foreshadows the question of insincerity, raising issues regarding the sincerity of tubers. The idea, however, of a "conspiracy theory," as it has been called, is missing the point, because it fails to account for the extreme popularity of the video and the success it continues to achieve.

Charlie bit me is an example of representation that aims at complete authenticity. The subjects are young children, thus not easily manipulated for fictional purposes. From a multimodal genre analysis, the video presents one *macrophase* and no significant transition (Baldry and Thibault 2006). Subjects are static and are shot in one close single take. No background noise or soundtrack is present. Both are seated in an armchair and the only speaking subject (Harry) looks at the webcam (his father), signalling a verbal and visual (i.e., eye contact) relationship with the camera. A vector analysis of Harry's gaze shows a simple pattern of alternation from his brother (Charlie) to the camera (his father). A vector analysis of Charlie's gaze is even more basic, in that he alternates gazes to his brother and father and to other unspecified points outside the frame. From a verbal standpoint, the video is equally static and barely verbal, presenting a series of very brief comments on the "action" from Harry. Such comments are exclusively referential and do not refer to anything but the here and now (i.e., *hic-et-nunc*), that is the elder brother bitten by the younger on his finger. A full transcription follows:

> Charlie: Auorr../owwah!
> Harry: Ah ah Chaarlie/Charlie bit me/ Ow/Ah/oohoo/oouch/
>   ouch//Charlie/OOUCH (cries)/Charlie.. that really hurts/
> Charlie: (laughs)
> Harry: Charlie bit me/and that really hurts Charlie/and it's still
>   hurting//
> Charlie: (grins)

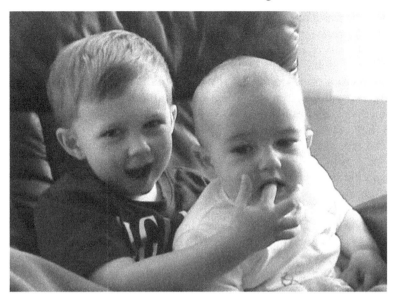

Figure 4.3 Screenshot from "Charlie bit me—again!"–Reprinted by kind permission of the Davies-Carr's www.youtube/HDCYT

A simple dynamic text such as this is nevertheless complemented by an extremely varied corpus of comments, which seems to define the interest of a great number of users in joining the discussion and share thoughts, values and judgments. The keywords "Charlie bit me" result in the striking number of almost 90,000 videos to date, which may be classified into different categories, for example *remakes* and *remixing*. Remakes are very popular on YouTube and exemplify further practices of self-representation. Remakes recreate an original video, turning it into something different, which nonetheless bear traces of the initial one. In this specific case, tubers play the role of the two kids, imitating their voices and body movements or using the kids' original voices, for example using dubbing. A second example is the remix practice, which used to be an alternative version of a song and that now includes a more comprehensive set of re-elaborations, which reprise shots or frames from the original video adding new visual or audio elements (songs, music, dubbing with or without *AutoTune*) or developing completely new creations, such as comics, parodies, caricatures or cartoons.

In the case of remakes, self-representation is realised thorough a playful transfer of identities towards the otherness of subjects who projected themselves in the remaking of the master-video text. In the case of remixes, the identity of the creator is realised in the created object itself, displaying discourse practices and multimodal and linguistic strategies employed in the representation and in the selection of the video to manipulate.

The cited examples allow a brief foray into the panorama generated around a homemade video realised for non-commercial purposes and whose goal was a private exchange of familiar moments with relatives living abroad. A private *mise-en-scène* may be originated by the authenticity of the relationships between the two brothers or set up by the father. It nonetheless turns into a global representation, interrogating and making explicit those textual relationships among interlocutors. In the next section, some linguistic aspects will be discussed using a manual and semi-manual analysis of comments carried out through Wordsmith, to investigate YouTube microtexts (i.e., verbal comments).

With regard to text selection, choosing YouTube comments for analysis may be highly controversial. Sweeping criticisms have been unleashed on this specific kind of web-based textuality. *The Guardian* in 2009 described users' comments on YouTube as follows: "juvenile, aggressive, misspelled, sexist, homophobic, swinging from raging at the contents of a video to providing a pointlessly detailed description followed by a LOL, YouTube comments are a hotbed of infantile debate and unashamed ignorance—with the occasional burst of wit shining through" (2009). In September 2008, *The Daily Telegraph* commented that YouTube was known for "some of the most confrontational and ill-formed comment exchanges on the internet" and reported on a new piece of software, *YouTube Comment Snob*, that hides comments posted on the video sharing community that "fail to meet a range of good English guidelines" (*The Daily Telegraph* 2008). A reporter from *The Huffington Post* noted in April 2012 that finding comments on YouTube that appear "offensive, stupid and crass" to the majority of people is very frequent. According to the reporter, although YouTube remains unscathed, the police and the Crown Prosecution service are acting on the principle expressed by the Department for Culture, Media and Sport that "what happens online should be equivalent to what happens offline" (Rundle 2012). However strong those criticisms may be, or perhaps as a result of these criticisms, YouTube's different forms of textuality have attracted the attention of scholars from different fields of studies (Adami 2009a, 2009b; Snickars and Vonderau 2009). As mentioned in Chapter 1, the fear of linguistic corruption and ultimate decay in web-based environments is very well represented in mainstream media and popular perception, and YouTube comments are no exception. A longitudinal analysis can also be particularly useful when analysing a transitional moment of popularity and "the hype around the platform," soon after YouTube was growing at an amazing rate of 75% per week, increasing the number of visitors of 297% from January to July 2006. By mid-2006, it had in fact 13 million visitors *every day* watching more than a hundred million videos (Cashmore 2006).

4.2.1   A Corpus of YouTube Comments: A Methodological Overview

The methodology used in this chapter is similar in some respects to that applied in Chapter 3. However, some steps have been modified for the purposes of this chapter and are listed here:

1. The creation of a dynamic monitor corpus of comments, providing the main dataset and access to YouTube textuality via one specific example, that is, the *Charlie bit me* video.
2. The use of software programmes (e.g., Wordsmith 5 and 6) to count frequencies and patterns of co-occurrence of linguistic features (mostly lexical) that have been selected as symptomatic of the written-spoken *continuum*.
3. The application of a corpus linguistics approach to study multivariate dimensions of language use and language variation in the corpus, with particular reference to lexical features, seen as symptomatic of the written/spoken variation (e.g., positive and negative keyness in comparison to some reference corpora, identification of frequent lexical bundles, observations on semantic preference and semantic prosody on some selected items).
4. A categorization of the comments into heuristic specimens and a discussion of the textual and multimodal relationships established between comments and between the video and its related comments.

4.2.1.1 Step 1: Corpus Design and Text Selection

As explained in the previous section, the selection of YouTube comments may be controversial, as mainstream media repeatedly report on their low quality in terms of bad or derogatory use of language, allowing the unleashing of verbal aggression by trolls (or haters), or the repetition of clichéd and banal formulae. However, further study is needed to test such general unfavourable reception with regard to a phenomenon that is still relatively unexplored from a linguistic and multimodal point of view. Such comments were selected over a 26-month time span to allow a longitudinal analysis, also allowing study of variation across time (2009–2012). The corpus contains less than one million words (828,900) and gathers posts written to comment on the *Charlie bit me* video, thus including some sixty-seven thousand comments (67,237). Unlike the corpora created in the previous chapters, the Charlie corpus (CC) only includes comments related to *one video*, focusing the study specifically on the micro-textuality generated by one very popular and viral non-commercial video. The *Charlie bit me* video rated seventh in the general top ten list of YouTube's most viewed videos ever as of January 2013, and is the most viewed, commented and rated non-commercial video to date, coming after Shakira's *Waka Waka* (*This Time for Africa*, rated sixth), LMFAO's *Party Rock Anthem* (featuring Lauren Bennett, fifth), Eminem's *Love the way you lie* (featuring Rihanna, fourth), Jennifer Lopez's *On the Floor* (featuring Pitbull, third), the YouTube celebrity Justin Bieber's *Baby* (featuring Ludacris, second) and Psy's *Gangnam Style* (rated first). Squeezed into a top ten list of popular hit and music stars, it is interesting to investigate what kind of popular reaction this video is stirring. As such, it may be assumed that it is a reliable example of how a very basic, barely verbal, if any, homemade and user-generated video clip can produce such massive *hypertrophic side-textuality*.

The corpus so far includes comments that have been collected longitudinally, starting from October 2009 and ending on January 2012, for a total of twenty-six months. The corpus aims to analyse a representative number of comments longitudinally and pave the way for further studies into forms of textuality, developing within other video sharing communities as well. In the case of LJC, a "time-frozen" analysis has been preferred, to capture a specific moment in time, whereas the corpus of video interactions explored in Chapter 2 was collected over a two-year time span (2009–2010, a shorter time span as video analysis is much more time-consuming, as it involves manual counting and fine-grained multimodal and interactional analyses).

The question of representativeness of CC has been considered on the lexical level, by "saturation" (Belica 1996) for specific linguistic features (e.g., lexicon size) of YouTube. To measure CC variation, the corpus has been divided into several segments of equal size, based on its tokens and has been *saturated* on a lexical level, because each new addition generated approximately the same number of new lexical items as the previous segment. In CC, linguistic features show very little variation, hence allowing the claim of representativeness for the corpus, at least on the lexical level. Saturation for representativeness has here been preferred to notions of balance, for similar reasons to those discussed in Chapter 3.

With regard to corpus balance, as CC is a specialized monitor corpus, including only comments to one video, balancing was not an issue, or at least not a significant one, as CC includes one kind of text referring to one context of situation. Comments were collected regularly each month and their maximum length determined by the platform affordances, that is, a maximum of 500 characters. Unlike video corpora and LJ corpora, the length of comments included in CC is determined by the medium. Considering that 500-character texts are the shortest texts analysed in this book, no minimum limit was established with regard to word number, and CC thus includes also one-word comments. Only comments written in English were incorporated and other languages excluded from CC.

Beyond the questions of representativeness and balance, there is also the issue of sampling units, which in this case are the comments. Demographic variables were almost impossible to collect and practically impossible to verify. An initial attempt was taken to build the corpus also in consideration of demographic variables, as in Chapter 3. Variables with regard to gender, age, language/s spoken and location were considered for 140 users, also comparing such data with similar sampling carried out for LJC, collecting data for 212 bloggers. However, a reliable report of such sampling is currently lacking. In spite of this, it soon became evident that YouTube user nicknames are, in the majority of cases neutral, devoid of gender specifications (e.g., FrozTftv, Big3019, qutipiecik, alphapudding119, BreadShuttle, WhiteWolfisawesome, TheRedParallel, etc.). Generalising, a significant transition along a cline has been observed in the domains of discourse identified in this book. In the chapter on videochat, participants were seen to be much more open to personal disclosure and concerned with *self-projection* that is typical of most online video interactions. In blogging environments, strong and opinionated blogs and blog entries often match the blogger's reluctance to self-disclosure, especially when it comes to fanfiction and with particular reference to the willingness to reveal one's own gender. On YouTube, participants are typically of two kinds: those who self-project their identity (e.g., the case of video uploaders, such as personal video responses or video blogs) and those who show themselves only as *verbal commentators* and, as such, are not willing to show anything but a neutral and undecipherable nickname. This claim is too general to boast general validity, but it is an observation that may help understand the usefulness of placing along a cline different variables in digital environments, as in the case of what may be defined as "willingness to self-disclosure." However, the small number of available data does not allow further comments on the question so far.

As a consequence to the difficulty of obtaining data with regard to YouTube commentators' demographic background, selection was carried out through *simple random sampling*, following CC design and purpose, that in this case could yield similar results to a stratified random sampling, in that it is based on homogenous groups (i.e., *strata*).

With regard to corpus size, the current number of comments, as of December 8, 2012, is 788,774, and the number of views is an astonishing 500,252,417. As the aim of collecting verbal comments was to study language variation across comments, the number of some 67,000 comments appears reasonable enough to justify a comparison to a general English reference corpus. As I have argued, no minimum comment length in terms of word number was set, as the extremely low number of characters allowed (i.e., 500 characters) would affect the reliability of the outcome. Consequently, entire micro-texts were included in CC, even though such texts may also be made up of one word and make both linguistic and multimodal analysis complex. To counter this problem, CC includes a reasonable amount of comments, to minimize the presence of one-word comments.

As in the study devoted to blog entries developed in the previous chapter, a context-governed sampling criterion was preferred to the less feasible and practicable demographic sampling. Akin to blog entries, the selected criterion was identified in the notion of popularity, which is also significant in the domain of discourse we are discussing. However, a proviso that must be made is related to a different kind of popularity that is being deployed in CC. In LJC, the collected entries are the most popular (i.e., commented on) of the LJ community in a specific moment in time. In CC, the collected entries are those written about a hugely popular non-commercial video. The CC corpus thus includes comments exclusively written in response to a video, defined as a *multimodal master-text*. The latter is defined as the video (or other multimodal artefact) that is the focal reference point for multimodal textuality or, to put the matter in slightly different terms, is the most salient resource. In other words, *a multimodal master-text is the most salient text in a given environment that is able to generate multimodal texts that refer back to it in a continuous cross-dissemination of multimodal hyper-intertextuality.* This concept will be discussed in more detail in section 4.3.

Unlike the previous chapters, the analysis of these comments has been focused on purely linguistic features. The corpus was built and tagged exclusively drawing on verbal language and putting aside the multimodal environment from which the micro-texts were extracted. This choice is not related to the fact that YouTube comments are mostly made up of verbal language: for example, the *Charlie bit me* video has generated thousands of written comments and far fewer response videos, that can be termed as multimodal comments. The reason for this is twofold: (1) multimodal comments (i.e. video responses) are dispersed on YouTube as they are, more often than not, not linked to the master-text; and, consequently, (2) multimodal comments do not share the same multimodal digital environment as directly as verbal comments do.

For these reasons, only verbal comments have been included in CC, because they belong to the same context of situation, being mapped across the same constellation of textual materials, directly related to the master-text. The purpose of such an operation will become clearer in sections 4.2.3, 4.3 and 4.3.4.

4.2.1.2 *Step 2: Corpus Tagging and Annotation*

As argued in previous chapters, addressing digital texts involves the consideration of several semiotic resources and the problem-oriented annotation has proven an efficient method to tackle specific research questions. Datasets contained in CC are quite similar to the verbal counterpart of LJC and have also been explored addressing research questions relevant to the aims of this book. As I have argued in the previous section, collecting verbal comments is helpful to investigate YouTube textuality, in particular to probe into spoken/written variation. Following the rationale adopted in the previous chapter, a corpus-based lexical analysis was applied to study language mode variation. However, in the previous chapter, the main corpus was annotated following two different criteria. The first one was purely multimodal, tagging and annotating the semiotic resources involved in the process of blog construction, also in the light of the development of specific language use, termed as blogEng. The second criterion was instead language-based and concerned with spoken and written variation across written blog entries, excluding comments. In this chapter, conversely, comments are the main domain of investigation. Such comments were collected following the rationale described in section 4.2.1.1.

One of the most significant research questions addressed in this book revolves around the notion of mode-switching, that may also be very broadly applied in observations within YouTube or other media sharing communities. However, considering that we have described this notion as *interactive and synchronous* by definition, that is occurring in synchronous modality and during the same communicative event, mode-switching *stricto sensu* is not typical of YouTube, at least not in the sense described so far.

It is true that many vloggers use YouTube as their blogging platform and that they may occasionally record a video of a videochat of their own or of other users. The best-known example is Shane Jensen, a former popular vlogger on YouTube, later banned from the video community. Shane Jensen is an American student who created and uploaded some controversial videos. Some of them reproduced his videochats with friends. His videochats showed typical patterns of interaction of video meaning-making events, including mode-switching. These videos were rather interesting, in that they added a further layer of complexity to the practices described in Chapter 2: a videochat was recorded and turned into a video clip, eventually uploaded on YouTube and, as such, *representing a representation*. The viewer was peeping into the two participants' video interaction, observing how they interacted and typically viewing the participant/recorder (e.g., Shane Jensen) from his back, but also seeing his projection on the second participant's (projected) screen. These interesting examples of double-layered video interactions are no longer available, so they have been excluded from analysis. However, these videos cannot be defined as specific examples of mode-switching, as they in fact only *reproduce* the original domain of discourse that in my model generate mode-switching, that is, as said, the videochat environment.

However, many examples of resource-switching can be found in video-based communities, akin to what happens in blogging communities. The alternation of semiotic resources within a broader framework of participatory and social digital interactions has been amply discussed in the previous chapter. In video sharing and social networking communities, a typical example of resource-switching is represented by a written, verbal comment followed by a video. Resource-switching is best represented in the chain of interactional exchanges in social networking websites, for example in *Facebook*, where comments, videos, photos, status updates and others are placed in inverted chronological order. All these are typical and very frequent instances of resource-switching and even a cursory look into these platforms allows assessment of frequency of such digitally constrained patterns of interaction.

With regard to CC construction, comments were collected following the criteria described in section 4.2.1.1, whereas annotation was carried out by deleting extra-textual features from the comments, for example author's nickname, date of publication, number of likes, dislikes, and so on. Despite the actual possibility of recovering the removed comments, it has been decided preliminarily that only those comments that were immediately available for reading should be part of the corpus. It was, in fact, deemed essential that comments that were still part of the interactional chain of asynchronous conversations should be extracted and inserted in the corpus, following the criterion of *availability*. However, some eventual findings show that such comments are rarely actually *read*. CC includes only available comments at the moment of collection.

In section 4.2.2, analyses will be carried out using overall CC as dataset. The study of spoken/written variation across written comments on YouTube has been developed using methods similar to those applied in the previous chapter. It cannot be overstated that different kinds of data call for different methods: videochat that have been studied in the Chapter 2 needed more fine-grained analysis, based on multimodal interactional analysis. An in-depth study of mode variation has been sidelined in favour of an analysis focused on the innovative interactional pattern of mode-switching.

Conversely, in Chapter 3 and in the present one, data are more similar, as similar are the research questions addressed. As said, mode-switching is by definition practically absent—or not so significant as in synchronous video-based interactions. Resource-switching is instead more consistent, as the chapter on blogEng has shown. Verbal datasets in both LJC and CC are, thus, mainly used to examine the written/spoken variation in digital communities. In section 4.2.2, verbal analyses into mode variation will be developed and details will be given with specific reference to CC. The linguistic features selected in the previous and in Chapter 1 will be followed accordingly to account for the spoken/written interplay in the case study.

4.2.2 A YouTube Case Study: The *Charlie bit me* Video Corpus-based Lexical Analysis

In this section, corpus-based lexical analyses will be used to study comments on YouTube. The perspective adopted is a longitudinal corpus-based analysis that allows an exploration of micro-textuality that has stirred attention in mainstream media, such as TV and newspapers, and on the web. Are YouTube comments so banal, trivial, repetitive and ridden with errors? Does *YouTube English* deserve such a poor reputation? Maybe the question is misrepresenting the issues involved: it is true that even a cursory reading of some casual YouTube comments reveal a non-standard and highly repetitive use of English. Misspellings, grammar mistakes, poor vocabulary, and repetitive and badly formed arguments are commonly found in comments in English. However, the book's main concern is not on how to improve English usage on the web, but to observe a linguistic trait and describe it from a situated perspective.

In this section, two main aspects will be tackled, with the aim of comparing the outcomes with those that have been emerging throughout the book, with specific reference to similar analyses reported in the previous chapter. The study of variation across the spoken/written *continuum* has been subdivided along four main lines, hence finding: (1) positive and negative keywords, in comparison to a general reference corpus, the ICE Great Britain corpus (see Appendix 3.1), to single out similarities and differences;

(2) positive and negative keywords in comparison to two subsets of the ICE corpus, namely the ICE spoken and the ICE written sub-components, to examine spoken/written variation more thoroughly and draw more delicate conclusions on both subcomponents; (3) most recurrent lexical bundles, to check whether they may be more closely associated with typical spoken or written bundles; and finally, (4) examples of semantic preference and semantic prosody.

As mentioned before, an MF/MD analysis will be partially applied to findings, to interpret data also under the light of a more fine-grained analysis, thus placing the identified features along the dimensions established by Biber (1988, 2003), and taking into account their positive or negative loadings, as well. However, variation will be mainly studied through keyness analysis and identifying lexical bundles, considering that using these methods can obtain results similar to an MF/MD analysis (Tribble 1999). For this chapter, Wordsmith 5 and 6 (Scott 2008, 2012) was used to carry out a keyness analysis to find the most prominent positive and negative keywords and associated features in CC. Prominence is computed with reference to a more general reference corpus and is a measure of statistical significance. Such features have been identified with the purpose of finding characteristics that are functional to the genre of YouTube comments. Keyness is also a measure of *aboutness* and, hence, it can provide some general indications as to what the genre might actually be about.

Variation has been computed modifying some of the standard default settings of Wordsmith, that is, a minimum frequency of 3 and a maximum of 16,000, considered a sufficient number for keywords as in the previous chapter. As an overview of CC, word length and a standardized type/token ratio were computed using the Wordlist function of Wordsmith. Table 4.1 reports on this and other statistical data with regard to CC.

Table 4.1   Data on CC

text file	CC_all.txt
file size	5,486,405
tokens (running words) in text	5,486,405
tokens (running words) used for word list	766,468
types (distinct words)	39,365
type/token ratio (TTR)	5.14
standardized TTR	34.01
standardized TTR standard deviation	65.82
standardized TTR basis	1,000
mean word length	4.37
word length standard deviation	3.15

The type/token ratio was standardized and frequencies normalized to a common 1,000-word basis to set a basis for comparability of different size files. Standardized type/token ratio is 34.01, which is considerably lower than in LJC (i.e., 43.96) and gives an initial indication of the degree of lexical variation within the corpus. Mean word length is quite similar to LJC. A preliminary analysis of the top 100 words in CC sorted by frequency showed that the most frequently recurring content items are: *video, Charlie, funny, cute, like, watch, love, baby, bit, YouTube, finger, people, kids, check, think, see, little* and *stupid*. Comparing these items with those from other wordlists (i.e., ICE, FLOB and BNC), CC includes a greater number of recurring content words, in particular some ascribable to a mere description of the video's contents, namely *Charlie, bit* and *finger*. Qualifying adjectives, expressing positive qualities, like *funny* and *cute* appear in the top 100 items (24th and 25th, respectively), whereas the first qualifying adjective describing a negative quality is *stupid* (92nd). Some specific kinds of content verbs are also very frequent in CC: *think, see* and *watch* are private verbs expressing mental states or non-observable intellectual acts, or, to describe them in systemic-functional terms, they are mental verbs, expressing emotion (*like, love*), cognition (*think, see*), and perception (*see, watch, check*). Another significant recurrent item is *YouTube*, making the name of the platform a visible and constant reference for commentators, quite unlike the case of LJ bloggers, who very rarely mentioned the community's name.

As shown in the previous chapter, CC displays a higher number of content items, showing a high level of *aboutness*, also implying a significant cross-referential activity on the part of commentators, who use words that can be easily placed in referential/descriptive categories. They describe or simply *reproduce* the contents that they have just seen on the video, but *verbally*, that is, using verbal language. This might be defined as a specific kind of mode-switching, but this new notion will be discussed in more detail in section 4.3.

Three additional wordlists have been used for the purposes of this chapter: (1) the one-million word ICE GB corpus (overall), (2) the ICE_spoken sub-component and (3) the ICE_written sub-component. For further details on the reference corpus, see Appendix 3.1.

No comparison between CC and other reference corpora, such as FLOB and BNC, was carried out, as was done in Chapter 3. A keyness analysis is reported in the Tables 4.2 and 4.3, comparing CC with the overall ICE GB and with its two sub-components, ICE_spoken and ICE_written as reference corpora. In order to create the keyword lists, the same wordlists used in the previous chapter were employed for the reference corpora, whereas the new wordlist created from CC was used to generate the keyword lists.

A comparison of the top ten positive items for keyness shows that the keyword list from the overall ICE-based list and from the spoken ICE-based list display exactly the same results, also in the same order. These two keyword lists are strikingly similar also in the top fifty positive items. They are

Table 4.2 Top ten positive keywords with three different reference corpora

No.	ICE as RC	ICE_spoken as RC	ICE_written as RC
1.	video	video	Charlie
2.	Charlie	Charlie	video
3.	cute	cute	funny
4.	funny	funny	cute
5.	views	views	views
6.	lol	lol	lol
7.	watch	watch	this
8.	this	this	I
9.	love	love	so
10.	YouTube	YouTube	watch

Table 4.3 Top ten negative keywords with three different reference corpora

No.	ICE as RC	ICE_spoken as RC	ICE_written as RC
1.	the	uh	the
2.	of	s	of
3.	s	of	to
4.	uh	uhm	in
5.	to	to	and
6.	in	that	a
7.	uhm	in	which
8.	a	and	as
9.	which	we	be
10.	that	yes	by

also quite similar to the keyword list created from the written ICE-based list, displaying eight items in common with the other two keyword lists. The most prominent positive items are very much situated in what can be expected to be the *aboutness* of the comments on the YouTube video as a genre: Charlie is the baby in the video and his proper name gives the title to the video. The eldest brother also repeats Charlie's name with remarkable frequency considering the short length of the video. Other positive items can be placed under the very general semantic heading of the video's specificity. It is a very popular video on YouTube and the fact that it is viewed frequently is attested by the positive prominence of items such as *watch* and *views*. It is also generally well received, as the positive prominence of positive qualifying adjectives *cute* and *funny* shows.

With regard to negative keywords, only three items appear in all three wordlists and, interestingly, all three are prepositions. *Of*, *to* and *in* all have negative loading along Biber's Dimension 1 (involved vs. informational production). The items that appear in the spoken ICE-based keyword list are negatively prominent in CC and are of mixed nature. For example, the interjections *uh* and *uhm* and *yes* are negatively prominent as they are features of typical spontaneous face-to-face conversations and are, as such, not likely to appear in written discourse, however informal and casual it might be. Non-phrasal coordination, such as *yes* and *yeah*, also displays positive loading along Dimension 1.

Contracted items or enclitics, such as *'s*, may also be indications of inconsistent transcription, but they also have positive loading along Dimension 1, as do first person pronouns, such as *we*. Another interesting item is *that*, rating 6th in the spoken ICE-based keyword list and 10th in the overall ICE-based keyword list. As mentioned in the previous chapter, *that*-clauses as verb complements, *that*-relative clauses and *that*-clauses as adjective complements have positive loading along Dimension 6 (i.e., online informational elaboration) in Biber's study on variation across speech and writing (1988, 2003). The fact that such features are negatively prominent would suggest a written-discourse orientation in YouTube comments. However, other items clearly show that some very typical items in the domain of written discourse are also negatively prominent in CC, for example *the* and *of*. We should remember at this point that the two most frequent words in FLOB and BNC World are *the* and *of*, and that both these corpora are mostly based on written discourse (FLOB is exclusively composed of, whereas BNC World mostly includes written data, cf. Chapter 3). *Of* as a preposition adds a negative loading to Dimension 1 (i.e., involved vs. informational production) identified by Biber's MF/MD analysis of the spoken/written variation in genre analysis (1988). Furthermore, we already mentioned the fact that Tribble (1999) also argues that *of* and *the* are usually associated with nouns, reporting on data that show that in academic prose *of* is used as a postmodifier in the N1 + *of* + N2 structure. The definite article *the* is also associated with nouns and in Biber's MF/MD analysis, nouns of the nominalization type are a feature with a positive loading along Dimension 3 (i.e., explicit vs. situation-dependent reference), while nouns of other types are a feature with a negative loading for Dimension 1 (involved vs. informational production). These features, thus, point to the negative prominence of such items in CC, similarly to those found on similar analyses on the LJ-based keyword list in the previous chapter. It thus may be concluded that some items typically associated with written domains of discourse are less prominent than expected in a corpus made up of written datasets. For example, *which* for sentence relative has positive loading along Dimension 1 (i.e., involved vs. informational production) and is in CC primarily used in *wh*-relative clauses and pied-piping constructions, which are salient features associated with Dimension 3 (i.e., explicit vs. situation-dependent reference). Returning to prepositions, *by* may signal by-passives and displays positive loading along Dimension 5 (i.e., abstract vs. non-abstract information). However, the prepositions *by* and *in* have negative loading along Dimension 1 (i.e., involved vs. informational production).

Concordancing both positive and negative prominent keywords has also shown that some features with positive loadings along Dimension 1 mark a reduced surface form, a generalized or uncertain presentation of information and a general fragmented production of the micro-texts that constitute the comments (Biber 1988: 106). Reduced surface form is marked, for example, by numerous *that*-deletion (e.g., *I think* [that] *Charlie is so cute*), contractions and enclitics (e.g., *'s*, *n't*, *'d*), *do* as an auxiliary (which substitutes a full verb phrase or clause), the third person pronoun *it*, first person possessive adjective *my* and demonstrative and indefinite pronouns. Dimension 1 is one of the most powerful factors in Biber's analysis and, as such, it is interesting to note that both positive and negative keywords have something to say on the matter. As argued in the previous chapter, positive keywords are the most frequent in comparison to a reference corpus: they show that some items occur more in CC than in a more general corpus of British English. Such data tell us something about the content (i.e., *aboutness*), but also about items that are more frequent. These items show that positive loadings along Dimension 1 are rather striking, especially for reduced surface form.

The negative prominent features, conversely, are revealing of what is less usual than might be expected. Such data are revealing for similar but opposed reasons. For example, the existential *there*, which is negatively prominent in CC, has positive loading along Dimension 6 (i.e., online informational elaboration) that is marked by co-occurrences of patterns typical of informal, unplanned type of discourse. Conversely, the subordination features, grouped under Dimension 6 and highly present in CC, mark informational elaboration "that is produced under strict real-time constraints, resulting in a fragmented presentation of information accomplished by tacking on additional dependent clauses, rather than an integrated presentation that packs information into fewer constructions containing more high-content words and phrases (as on Factor 1)" (Biber 1988: 113).

In the top 300 positive keyword features, private verbs that also have positive loading along Dimension 1 are: *believe, feel, guess, hope, imagine, know, mean, see, show, think*. As argued, these are all mental verbs in Halliday's terms (Halliday and Matthiessen 2004) of various kinds and all imply a mental process with a human agent as *senser*. In the CC wordlist, items related to the digital domain, such as those listed in Chapter 1 (see Table 1.2), include reduplicate punctuation marks (e.g., *what???!!!*), reduplicate phonemes (e.g., *whaaaaaat?, goooood!, soooooooo*), misspelling/non-standard spelling (e.g., *realy*), capitalization (e.g., *CHARLIE*), lower case (e.g., *i think im right*), acronyms, shortenings, abbreviations (e.g., *omg, pls, app, btw, thx, lol, lolz, rofl*), onomatopoeias (e.g., *arrgh, AHAHAHA, OUCH*). Note that the latter are usually also associated with capitalization.

Akin to the findings on LJC, a negative keyword list generated with the overall ICE-based list as reference corpus displays many mixed and hybrid negatively prominent features. It cannot be overstated that such data are only partially revealing in that they compare British English with global varieties, including the variety of *English as a Foreign Language* or *English as a Lingua Franca*. As I have argued previously, a reliable account of the language used by YouTube commentators is virtually impossible.

With regard to the consistency of findings across time, a keyness analysis has been carried out using 26 subcorpora, (CC1, CC2, CC3, CC4, CC5 and so on), one for each month of the longitudinal analysis and using only the overall ICE-based list as a reference corpus. Data show that comments are practically stable across the selected time range. The first top 10 items were analysed for each of the 26 subcorpora with at least 6 common items in the 26 positive keyword lists. 13 sublists yield the same top 10 items, that is, the same as shown in Table 4.2, whereas the remaining half displays a maximum of 9 and a minimum of 6 common items.

A sample analysis of the five-word clusters shows some interesting recurring patterns. With regard to keyword and keyword linking types, the top 15 lexical keywords have been reported, sorted by frequency, from the overall ICE-based list with the 10 most-linked types. Number of links and total hits are also reported. The collocational span is established by default, with five words to left and right of the node. A complete list of all linked keywords shows that a significant number of foreign words (e.g., German, Italian, French, Spanish) are included in the corpus, because they were inserted in partly-English comments, so they have been excluded during corpus tagging and annotation. In Table 4.4, many of the features listed by Danet (2010) as instances of English computer-mediated communication are found in the linked types. Recurring features include misspelled words, lowercase and capitalization, swear and taboo words, nicknames, onomatopoeias, acronyms, contracted words and enclitics, single letters of uncertain interpretation and words without vowels, similar to what is commonly found in text messages.

A comparison with a similar analysis in the previous chapter shows that CC yields many more results, as CC is considerably bigger than LJC. We have accordingly slightly increased the number of lexical keywords from 10 to 15, whereas the linked types have been kept constant in this section. The top 15 keyword plus link types have been reproduced here to provide a list of the most common patterns of link types sorted by frequency, including hits per sorted keyword.

In section 4.2.3, a study into the most frequent lexical bundles will be carried out to explore how words combine in YouTube comments, with the final goal of starting a functional categorization of comments, which will be discussed in section 4.3.

Table 4.4 Keyword links in CC

Keyword	Links	Hits	10 most-linked keywords
video	196	8,305	they're, brothers, rage, song, dislike, hilarious, disliked, freaking, beat, tht
charlie	164	7,509	accents, most, crap, dumb, brother, please, watch, thumbs, little, evil
funny	159	6,402	evil, soo, you're, search, fucking, charlie's, sweet, soooo, he's, funnier
like	156	3,909	lool, cutest, soo, plz, million, utube, damn, cant, sooooooo, bite
cute	126	6,258	like, damn, harry, many, get, charlie, guys, please, i'm, cant
views	121	5,280	why, like, super, i'm, charley, awww, click, chubby, lot, funniest
watch	114	3,574	videos, love, popular, chubby, clicks, soo, button, get, people, channel
finger	103	2,292	mil, wow, please, bit, idiot, stupid, soooooo, annoying, clip, dude
love	87	3,024	holy, twinkie, it's, aww, cant, guy, ugly, cool, biting, check
bit	85	2,370	luv, like, mouth, didnt, wow, rofl, shit, me, cutest, laughing
baby	84	2,303	aint, thats, funny, gotta, viewers, hey, bite, gump, bet, ugly
people	84	2,221	bet, wow, crazy, fucking, check, nice, mouth, boring, its, love
kids	81	1,834	yum, giggle, hate, im, wierd, isnt, hell, watch, fuckin, stupid
Youtube	80	2,296	cant, adorable, super, dislike, bitten, wierd, they're, baby, stupid, me
little	69	1,260	utube, ever, thx, dude, lmao, wtf, laugh, baby, hurting, u

4.2.3   Words in Combination: Lexical Bundles on YouTube

In this part of our analysis, Wordsmith 6 (Scott 2012) was used, changing the setting for cluster computation within CC. The wordlist previously created for a keyword study was also used to create an *index*. Settings include all words that form clusters from a selected cluster size from four to five words, minimum frequency of 30 occurrences and maximum frequency of 50 percent, with stop at sentence break. The top 32 four-word and five-word lexical bundles are reported below, with the number of occurrences in brackets. The number of reported bundles includes frequencies greater than or equal to 0.01 in CC. Frequencies inferior to 0.01 have not been reported, but both the four-word and five-word lists were manually analysed. Bundle frequency is of 0.02 from number 1 to number 13 of the list, whereas from number 14 to number 32, frequency amounts to 0.01:

1. finger in his mouth (253)
2. the most viewed video (253)
3. Charlie bit my finger (252)
4. I love this video (248)
5. yum yum yum yum (244)
6. yum yum yum yum yum (242)
7. this is the most (203)
8. babies babies babies babies (189)
9. babies babies babies babies babies (186)
10. Charlie that really hurt (182)
11. thumbs up if you (181)
12. most viewed video on (175)
13. subscribe subscribe subscribe subscribe (168)
14. viewed video on YouTube (164)
15. sweet sweet sweet sweet (160)
16. most viewed video on YouTube (159)
17. subscribe subscribe subscribe subscribe subscribe (159)
18. sweet sweet sweet sweet sweet (158)
19. they are so cute (142)
20. the most viewed video on (137)
21. put his finger in (134)
22. this video is so (134)
23. has so many views (131)
24. I like this video (128)
25. get so many views (127)
26. Charlie Charlie Charlie Charlie (120)
27. Charlie Charlie Charlie Charlie Charlie (117)
28. is this the most (117)
29. the most watched video (117)

30. this is so cute (116)
31. this is so funny (116)
32. your finger in his (110)

This list is quite instructive and even a quick examination of the most frequent bundles in CC allows specific considerations. First of all, it is not surprising that four-word bundles are more frequent than five-word bundles. Furthermore, the degree of *aboutness* in the bundle list is apparent and several bundles can be grouped in similar semantic categories, whereas others (i.e., 10, almost one-third) are reduplicative in nature, as they present the same word (or onomatopoeia, such as *yum*), which is repeated four or five times, forming a particular kind of bundle. It is quite striking that the reduplicative bundles are constant across time, as they have been found in various subcorpora and are not typical of specific moments in time. With regard to other bundles, we may well argue that some short chunks of language (ranging from 16 to 50 characters, out of the 500 allowed) are revealing with regard to some tendencies in CC. Examples, such as *finger in his mouth, Charlie bit my finger, Charlie that really hurt, put his finger in, Charlie Charlie Charlie Charlie (Charlie)* and *your finger in his*, clearly display a very high referential content, as *they all refer back to the video verbally*, hinting at the protagonist's name or at the salient features of the micro-story, including content items such as *finger, hurt* and *mouth*.

Another group of items shows a different focus, which is not based on the video referential content, but on the video itself, on a sort of meta-reflexive stance on its popularity. Bundles, such as *the most viewed video, most viewed video on, viewed video on YouTube, most viewed video on YouTube, has so many views, get so many views, is this the most* and *the most watched video*, at the same time show the commentators' surprise (both in positive and negative terms) at the video's huge popularity. These bundles are fairly common in prose; as Biber et al. claim (1999: 1023), they "have copula *be* as the main verb, and either the demonstrative pronoun *this*, or existential *there*, as subject." Examples include *this is so cute/funny, this is the most, this video is so.* Also noun phrases with *of*-phrase fragments or with other post-modifier fragments are typical of prose and can be ascribed to this group.

A third group adds to an evaluative stance, with examples such as *I love this video, thumbs up if you, sweet sweet sweet sweet (sweet), they are so cute, this video is so, I like this video, this is so cute* and *this is so funny*. Other less common examples may be placed in two different groups and these are *this is the most* and *this video is so*. When the former collocates with verbs such as *viewed, clicked* and *watched*, it clearly refers to users that comment on the video's popularity, whereas when it collocates with qualifying adjectives, like *cutest, adorable, prettier, stupid, popular, overrated, liked, rated* and so on, it is part of the evaluative stance group. In the case of the *this is the most* bundle, however, there are many more concordancing lines indicating that the most frequent use is within the "Questioning popularity" grouping, that will be discussed in section 4.3. The latter bundle collocates with *viewed* and *watched* most frequently (see section 4.3 for a full description of functional groupings of YouTube comments). With regard to the second problematic bundle, that is, *this video is so*, the concordancing function for this bundle of Wordsmith has revealed that this can be placed exclusively in the evaluative stance group. After concordancing it, the bundle was re-sorted by R1, showing both qualifying positive and negative adjectives in R1 position, such as *cute, funny, stupid, fake, genial, great, pretty, old, popular, sweet, innocent, appealing, viral, weird* and *dumb*. In some cases, adjectives are in R2 position, whereas a pre-modifier adverbial is placed in R1 position, which is, in the over-whelming majority of cases, *fucking* and *damn*. The most prevalent type of lexical type in conversation is, according to Biber et al. (1999: 1002), a clause fragment, consisting of a personal pronoun subject followed by a verb phrase. In many cases, Biber et al. claim that the verb phrase is ex-tended by the beginning of a following complement clause. These types of bundles are very frequently found in CC, as this third group attests, with examples such as *I love/like this video*.

The reduplicative group includes a significant number of bundles, the largest of the selected sample from our dataset. Concordancing one sample reduplicative bundle, that is, the repetition of the baby's name, *Charlie*, in-terestingly reveals larger bundles, with *Charlie* in capital letters as the node and other occurrences of the same item in L5, L4, L3, L2, L1 and R1, R2, R3, R4 and R5 positions (five items on the left of the node and five items on the right of the node). The same may be said about the first reduplica-tive bundle sorted by frequency, *yum*, that repeats the same item even from L7 to R7, increasing the bundle to an astonishing 15 words. However, the classification of mere repeated words as bundles is highly controversial, as it cannot be argued that they are words that co-occur. They are merely added to one another as mechanical additions and are not meaning-making bundles. Biber et al. (1999: 1014) claim that a few bundles in conversation do not fit into any of the other categories, and report as examples: *no no no no, two and a half, three and a half* as four-word bundles and *on and on and on* as five-word bundle. They suggest that apart from numeric phrases, other bundles are used for emphatic negation or for marking the continua-tion of some process. The last category of conversational bundles is defined as "meaningless sound bundles" and includes *mmm*, used to backchannel for agreement or affirmation and also repetitive sound bundles, typically used for musical purposes. It would be difficult to consider reduplicative bundles from our dataset as "meaningless sound bundles," as they are, more often than not, content words from different word classes, such as noun, adjective and verb, for example, *Charlie, sweet* and *subscribe*. They may also be loosely placed within the above-mentioned categories, but they can be more closely associated with conversation than with prose, let alone academic prose. It is difficult to imagine such an extensive use of reduplicative bundles in other environments and LJC did not yield similar results, not even in bundles of less than four words. These are used for emphatic purposes and play a significant role as far as their frequency is concerned.

It is nonetheless quite evident from this discussion that findings from an analysis into CC is very different from similar analyses carried out on LJC. Bundles are here rather topical, referring back to the master-(video)text and the most common bundles in both speech and writing are, quite strikingly, *less* frequent than expected. From some observations in this section, it emerges that some bundles, more common in general corpora, are present, but are significantly lower in number in CC than in LJC. Words that co-occur in CC are lexically salient, in the sense that they are semantically linked to the text they refer back to (i.e., the video), even though the very notion of "semantics" needs to be interpreted in very broad terms. Most four-word and five-word bundles are immediately recognizable as belonging not only to a specific video (it is almost impossible to forget we are immersed in the Charlie experience!), but to a specific platform, with its rules and linguistic habits, such as, to name but the most striking, the extensive use of reduplicative bundles.

Further study is needed to assess the nature and significance of this specific kind of word combination on YouTube, as reduplicative bundles are a significant part of this specific textuality and cannot be dismissed as mere "nonsense." Possible research questions include the exploration of commentators' motivations in reduplicating words: are they trying to get other people's attention? Are they trying to be *actually* read? As only a handful of comments, published in inverted chronological order, is visible in the video's main page, the chance of these comments being actually read is very low. This is especially the case when a video is popular, as the page is continuously refreshed and thus changed by and through the addition of new comments. However, popularity is the reason why so many users want to comment. Their reasons may also be unrelated to their personal experience of viewing the video, as they may well be simply trying to promote their own channel by posting other kinds of messages in the hope of being read. In this sense, the paradox lies in the fact that while users try to get their messages across by exploiting the video's popularity, it is also thanks to such popularity that their *chances of being read are actually debunked.*

The phenomenon of deviating or publishing materials that are external or even indifferent to the video they are related to is very well known under the name of *spamming*. From a linguistic point of view, spamming can be considered as a conscious break from what is conventionally known as "rules of communication." If we postulate that YouTube comments are ongoing web-based and asynchronous conversations, they consequently break, in many ways, Grice's conversational maxims (1952, 1989). Adami (2009a,

2009b) has amply discussed the fact that YouTube comments in general break Grice's relevance maxim, as they are not related to one another. We will discuss this aspect in more detail in sections 4.3 and 4.4.

4.2.4 Semantic Preference and Semantic Prosody in CC

Other concepts that will be taken into account in this chapter are the notions of *semantic preference* and *semantic prosody*. Semantic preference is one of four types of semantic relations identified by Sinclair (1996, 1998) and depends on the relationship between a set of frequently occurring collocates and shared semantic properties. It has been defined by Stubbs (2001: 65) as "the relation, not between individual words, but between a lemma or word-form and a set of semantically related words." In semantic preference studies, research shows that some word forms, or lemmas, tend to co-occur primarily with words belonging to a *restricted set* of semantic categories. For example, the verb *undergo* is usually followed by verbs of change, training or testing and preceded by items that indicate involuntariness, thus this verb has a semantic preference for words denoting change (Stubbs 2001).

As Moreno argues (2009: 242), Stubbs' definition draws a clear difference between collocation and semantic preference, as the former is found in combinations of actual lexical units, whereas the latter implies a higher level of abstraction, as it involves the combination of one word with a group of words that share some semantic features. Among the most remarkable examples, *large* is often associated with terms of quantity or size, such as *number*, *scale*, *part*, *amounts* and *quantities* (Stubbs 2001). Another example is represented by intensifying adverbials, such as *utterly*, *totally*, *completely* or *entirely* that are preferentially associated with words that mean a lack or status change, such as *irrelevant*, *unexpected*, *altered* or *different* (Partington 2004). As several studies claim (Partington 1998, 2004; Sinclair 2003; Hoey 2004, 2005; Moreno 2009), even though a word displays the semantic preference to co-occur with one or more groups of lexical units, these associations are neither exclusive nor mandatory. Language always tends to be constructed in pre-fabricated lexical units, but they are not fixed and frozen, as language exhibits a degree of flexibility. As Partington argues (2004: 146), "language users are able to swim against the current—can 'switch off' primings—when they seek particular creative effects. Semantic prosody and preference do not ordain that counter example *cannot* happen, just that they *seldom* happen." As Moreno explains (2009: 243), the Sinclair school has not established a distinction between a base and its collocates, as in the phraseological tradition. In corpus linguistics, the node can be *any* word that is under study. This is why semantic preference is not studied in terms of base and collocates, but only in terms of words, or nodes, despite its phraseological position. In effect it is easy to understand that we find instances of semantic preference in terms of both node and collocates, even though it is mostly found in nodes.

Link types, as reported in the previous section, give an indication of main collocates for the most frequent items in CC. However, some content words were selected for further analysis to shed light on semantic preference in CC. This study is not interested in general semantic preference for the selected words, but only showing examples of semantic preference within CC.

With regard to the noun *video*, we find that it tends to co-occur with *most viewed* or *most watched*, thus referring to all those comments that make reference to the video's popularity, in either positive or negative terms. Another set of preferential collocates is made up of a series of both positive and negative adjectives, such as *funny, cute, stupid, popular, great, new, good, cool, old, nice, awesome, hilarious, gay, favourite, amazing, boring, little, famous, retarded, dumb, adorable, viral, sweet, simple, British, rated, pretty, overrated, classic*, as the first 400 collocations reveal.

Analysing the semantic preference of the proper noun *Charlie*, we find that it co-occurs with two main patterns: the first more or less repeats the words uttered in the video, with collocations such as *Charlie bit/eats/ate/me/ my finger* or *Charlie (that/'s) really hurt/s*, whereas the second is descriptive, with the pattern of copula *be* plus adjective, occasionally preceded by pre-modifiers (e.g., *apparently*) or followed by post-modifiers (e.g., *just*) or with the less frequent descriptive verbs, such as *looks like, seems, sounds, starts, giggles, laughs*.

With regard to semantic prosody, Stubbs notes that this usually occurs at a higher level of abstraction and that semantic preferences are typically more restricted in size and semantic categories. To return to his example *undergo*, he argues that the ensemble of preponderance of words preceding the verb indicating involuntariness and of words following the verb indicating change, coupled with the idea that these notions (i.e., involuntariness and change) are usually deemed undesirable, lends negative connotations to constructions involving *undergo*. This negativity around constructions with *undergo* represents its negative semantic prosody. Both Stubbs (2001) and Partington (2004) insist on generality as a main feature of semantic prosody, although the latter very frequently informs a construction in positive or negative terms.

An interesting example from CC can shed light, in very broad and perhaps far-fetched terms, on the semantic prosody of a lexical item, that comes as the 43rd occurrence, sorted by frequency. It is the noun *accent*, reporting on a hotly debated topic with regard to the video, that of Harry's accent (the eldest and only speaking brother). Accent is mainly described as having strongly positive or negative associations, as concordancing also shows: *adorable, annoying, awesome, beautiful, cute, funny, hilarious, lovely, nice, posh, weird, wicked* and *wonderful*, but with a very high correlation with evaluations of the accent's geographical collocation, with L1 overwhelmingly being *British, English* and, occasionally and ironically, *American, Scottish* and *Australian*, or, more specifically, enquiring about the exact geographical collocation, with comments asking to specify the kids' provenance: for example *Powys, Shropshire or Hertfordshire?*

If we consider L1 positions, when a pattern such as *the/his/their/this/ that/your* appears before the node *accent*, the most frequent preceding items (L2–L3) are *I love/like/adore*. Taking for granted that only one node has been always selected in this symptomatic analysis, it may be assumed that commentators take up polarized evaluative stances in these examples. Semantic prosody of the selected items clearly shows that commentators express either positive or negative attitudes in judging the video or its contents. To put the matter in simple terms, the cited qualifying adjectives that are most commonly associated to the selected (and very frequent items) are strongly positive or negative, with a low degree of neutral stance. A wealth of comments is devoted to the boy's accent, stirring a hot debate on the prestige of British English that is apparently still widespread. In this case, accent is a metonym for oral discourse that is barely verbal on the video. Despite the fact that comments come from the Anglophone world on a global scale, users verbally exhibit a persistent bond with linguistic ideologies, describing British English as *superior*, *more beautiful*, *more musical*. An analysis of eight item bundles displays an almost unanimous appreciation of the British accent used in the video.

A plethora of comments is thus occasioned by a banal interaction involving two children. However, a very high level of repetition is instantiated in such comments as shown in the analysis of CC. Due to repetition and frequent lexical patterns that display the *aboutness* of CC, some tentative categorizations can and will be attempted in the following section.

4.3   A MULTIMODAL RELEVANCE MAXIM

As shown in several studies (Adami 2009a, 2009b; Andrejevic 2009; Kessler and Schäfer 2009; Lange 2009; McDonald 2009; Vonderau 2009), the degree of interaction is actually very low on YouTube, because dialogue between users—in those rare cases in which it is present—is limited to the space of a visualized page, as mentioned in the previous sections. In other words, dialogues cannot survive the very brief time lapse of a web page, because almost no user is interested in reading and engaging in virtual conversations beyond the web page that is visualized at the moment he/she is logged in. These texts can be defined as a series of mutually exclusive micro-texts, despite their apparent interconnected relationship. Their textual relationship with what comes after and what comes before is thus relatively low, showing a consequent low overall cohesion and coherence, as is also evidenced by textual analysis. However, what seems apparent from these preliminary linguistic observations is that comments do not function as "dialogues" and they do not display the typical features of intertextuality. As I have already mentioned, what seems evident is the establishment of a relationship not between comments, but within the comments and the foregrounded video, defined in this chapter as a master-text. The latter is interpreted as a reference point, as the "source" of intertextuality, as most comments directly or indirectly refer back to it. If we follow Adami's contention (2009b) that YouTube comments break Grice's relevance maxim (1952, 1989), we do not take into account the multimodal interweaving of semiotic resources that make up the multimodal environment of video sharing communities. It may in fact be assumed that a *multimodal relevance maxim* is at work. Comments need to be consistent with the main communicative focus of multimodal interaction and the most salient semiotic resource: the foregrounded video. A multimodal relevance maxim can be useful when attempting to adjust well-established linguistic notions to digital domains of discourse.

YouTube is a complex multimodal environment, encompassing old and new textual practices and conversational habits at the same time. The system is always the same, in that the website presents a logo top left, then a centre-top search engine, where users can type keywords or topics of interest to be searched in the website and the video in the foreground, with a top-left arrangement, in tune with Western literacy (left is the most and first viewed element in a page or screen, as we read from left to right, cf. Thibault 2000).

A functional interpretation of the data reported on in the previous section allows the categorization of comments into a series of groups. In a previous study (Sindoni 2011c), I proposed a taxonomy using a smaller dataset, but in the present analysis, the notion of comment groupings is preferred for two main reasons: (1) a taxonomy (or categorization) is not flexible enough for a digital web-based environment such as YouTube, whereas the notion of grouping implies a less rigid categorization; and (2) the current findings only refer to *one* video and this does not allow for a broader application of the outcomes discussed in this book.

These comments thus represent isolated micro-texts, which apparently do not relate to each other and present a low consistency across exchanges.

Table 4.5   Groupings of YouTube comments, referred to the "Charlie bit me" video

Quotation	Report words or whole utterances from the discourse produced in the video
Evaluation	Express judgments in forms of approval or disapproval on the video or parts of its contents
Questioning popularity	Challenge popularity, number of views, etc.
Self-promotion	Promote the view of one's own or other people's video (including spamming)
Deviation	Interrupt coherence in the chain of comments with completely unrelated observations (including spamming)
Removal	Text removed. The only available information is author and date of publication

However, as in the previous sections, we may compress this huge number of comments into a limited number of functional categories, shown in Table 4.5.

Groupings may be considered as crude indications and some of those shown above, in particular "Self-promotion" and "Deviation," overlap. More specifically, some of the comments in the "Self-promotion" grouping may also be included in "Deviation," because self-promotion is a deviation and an interruption in the coherence of topic threads treated in the comments. Furthermore, spamming may be both "Self-promotion" and "Deviation" at the same time, whereas not all comments that deviate from the *commenting on the master-text activity* should be considered "Self-promotion."

"Quotation" are more or less direct quotes from the video, reporting all or parts of the utterances or the children' names; in this sense, they project from the video, creating paratactic quoting in Halliday and Matthiessen's terms (2004). This is an example of how Halliday's language-oriented model can be applied to other semiotic systems, as, in this case, projection does not involve two verbal clauses, but two different semiotic systems, that is, a visual resource, the video, and a verbal resource, the written comment (Sindoni 2011b). This is the most significant grouping in terms of frequency. Users seem to be very keen on capturing some items from the master-text and merely copying it onto their own comments, as the high frequency of items such as *video*, *Charlie*, *bit*, *finger*, *that*, *really* and *hurt* attest.

An epiphenomenon generated by the repeated views of this specific master-text is represented by constant forms of commenting about viewing the video. This particular grouping can also be interpreted as a multimodal instantiation of the oral mode into the written mode. It is a type of asynchronous mode-switching, whereas the master-text features some basic child's utterances (no exchange or interaction is actually instantiated in the video), the related comments mode-switch asynchronically and turn the barely spoken verbal into something fully verbal and written in mode. However, the "copying and pasting" technique is here used to semantically change the verbal mode instantiated first by the master-text. These observations will be also taken up in our conclusions.

"Evaluation" comments are mostly relational attributive clauses in Halliday's experiential analysis (e.g., "This/the video is. . ."), but may be analysed also in interpersonal terms (Halliday and Matthiessen 2004), as modality is used to express writers' positions, opinions and stance. Incidentally, evaluation is also a fundamental in appraisal theory (Martin and White 2005).

The significant number of comments written by the YouTube "haters" sub-community, an important sub-grouping, points to those who comment on videos and/or other people's comments in negative or derogatory terms. "Evaluation" comments may also overlap with either "Questioning popularity" or "Quotation" or both. "Questioning popularity" comments basically ask questions regarding the popularity of the video and may be explored using different criteria. Comments pertaining to this grouping are rather paradoxical in that commentators challenge the number of views, while at the same time contributing to the total number of views! For example, a comment like "Why so many people comment on this video?" is paradoxical, because it is commenting precisely on what it is challenging. In this sense, this kind of comment is a *meta-comment*—a comment on comments.

Such groupings, though rather unstable, may prove to be useful working hypotheses. Examples of each grouping are reported here:

Quotation:

- adsoiderful: and that really hurt
- Sternenmacht: Charly bit me aaaaaaaaaaaaaahhhhhhhhhhhhhhh? XD
- 19jesse97: Charlie bit me
- Vidioinspector: charlie! that rly hurt!
- 85Jinx: it really hurts charlie and its still hurting (LMAO) !!!!!

Evaluation:

- Hola24143: Cute <3
- BethJunginger: haha . . . stupid but cute
- KREJZYGUY: why does charlie remind me so bad of Britney Spears xD
- lizzlovesD: Never gets old makes me laugh everytime lol
- 2zach3: the kid has more teeth than hair!!!

Questioning Popularity:

- ostepobss: its pretty incredible that this video can get that many views
- dudelittle2: 240 000 000 MILLION VIEWS. EIGHT times the entire population of canada. HOLY MOLY
- LpKing91: mhmm.not trying to mean but how in the f*** does a british kid yelling "owwwwwwwww Charlie bit me" get so popular and Viral like this??
- TheParodiesShow: HOW IS THIS THE MOST POPULAR VIDEO ON YOUTUBE??????
- PSTSkatecrew: WTF! WHY OVER 240 MILLION PPL WATCH A BABY BITE ANOTHER BABY!?

Self-promotion:

- isaiahman777: PEOPLE PLZ SUB I AM TRYING TO BUILD MY CHANEL

- FriedPiggy6000: can somone subscribe to me and i swear ill subscribe back if u do to me . . . thanks and please subscribe to meh
- HBTATheShow: check my page for jackass like vids!
- fruitloop33344455555: check out my channel PLEASE i need more views than my friend:P Shes Better Than ME–so i need more views She is WINNING right now so HELP MEHH!!!
- franxz24: everybody go to the video Matt112417 singing kum by yah

Deviation:

- gingboyz55: i like waffles
- Powerman341: Israel Kamakawiwo'ole
- MrPtrk19: okay everyone needs to read this know there is a YouTuber and he is going around spamming everyone. he had spammed me too many times so i suggest that you block him KNOW his YouTube user name is robertjohnson487–I suggest you do this
- MrQwertyui1239: You may not want to be a Christian because it's boring or hard to believe or you want to do things your way. Well whatever the reason, I urge you to go to Jesus. I for one thought being a Christian was what I said above but the Lord showed me miracles that changed that. Jesus is the only way to heaven and without him you have nothing to look forward to when you die. I'm not forcing you to be a Christian cause God doesn't force you. It your choice. May God lead you to
- Naruto10000Fan: DONT READ THIS CAUSE IT REALLY WORKS, YOU WILL GET KISSED ON THE NEAREST POSSIBLE FRIDAY BY THE LOVE OF YOUR LIFE, TOMORROW WILL BE THE BEST DAY OF YOUR LIFE, HOWEVER IF YOU DON'T POST THIS COMMENT TO OVER 3 VIDEOS OR YOU WILL DIE WITHIN 2 DAYS, NOW THAT YOU STARTED READING THIS DONT STOP, THIS IS SO SCARY,POST THIS TO OVER 5 VIDEOS IN 143 MINUTES WHEN YOUR DONE PRESS F6 AND YOUR CRUSHES NAME WILL APPEAR ON THE SCREEN IN BIG LETTERS, THIS IS SO SCARY BUT REALLY WORKS

With reference to the multimodal relevance maxim, items such as *video*, *YouTube*, *views*, *finger*, *bit*, *baby*, *kid*, and *mouth*, and related lexical bundles, such as those mentioned in section 4.2.3, contribute to feed the semantic network of "Quotation." The prevalence of "Quotation" comments also conjures up a tendency towards relevance, that is deployed in the stable reference to the master-text across CC, which gathers comments in longitudinal fashion. However, it cannot be overstated that a huge amount of comments are oriented in precisely the opposite direction.

Spam comments have been grouped under the general headings of "Self-promotion" (i.e., directing users towards specific videos/channels with the aim of advertising them) and of "Deviation," which is a less restricted grouping for spamming, including all those comments that contain completely unrelated topics, but excluding "Self-promotion" spamming. Spam comments thus display the tendency towards what may be termed as the opposite direction of relevance, that can be defined very broadly as *deviation*. To put the matter in simple terms, comments in CC can be very broadly placed in two functional macro-categories in terms of communicative purposes: *the first is oriented towards relevance, the second towards deviation*.

It may be noted in passing that the use of adjectives is rather repetitive and so scarcely varied as to allow some hypotheses, such as a presumed "universality" in the reception of such a video featuring a widely shared situation in all cultures (e.g., feelings of endearment towards children). Other hypotheses are more concerned with this specific text type, which features a reiteration of words written by preceding commentators, to be interpreted as a passive reproduction of others' evaluations. Alternatively, commentators appear to master a low range of registers, as they who seem unwilling (or unable) to employ differentiating linguistic strategies (e.g., use of synonyms). It is difficult, perhaps impossible, to establish whether this is due to a lack of linguistic tools or the willingness to use them. These observations provide a somewhat limited, but emblematic, scenario of textuality and metatextuality in YouTube. Although the master-videotext is short and simple in its structure, topic and realization, the related corpus of comments is quantitatively extremely rich and qualitatively rather poor.

This consideration does not imply an evaluative judgment on the presumed low quality of comments, but highlights the possibility of liable groupings which could allow some generalizations, with the goals to make textual exploration possible, also in the light of the comments' embedding in a multimodal environment. This undertaking does not require a rigid taxonomy, but invokes a viable foray into a still little known textual world.

4.4 SELF-REPRESENTATION AND IDENTITY: VERBAL, VISUAL OR MULTIMODAL?

This chapter has investigated textual phenomena on YouTube from different standpoints, indicating different but compatible paths of enquiry. The representation of identity on YouTube videos endorses different conceptions of the self, which, as discussed, are bound to personal and collective ideas about privacy and protection of privacy from outer attacks (e.g., spam and stalking). Different settings for privacy purposes correspond to a different vision of visual, textual and metaphorical space of self-representation. How self-representation is instantiated in comments has also been the focus of this chapter, also in the light of the interrelationship between the oral mode (i.e., audiovisual video clips on YouTube) and written comments. Videos have also been subdivided into two main functional sets: (1) *master-text*, that is the foregrounded video, both in visual and textual terms (i.e., the video that is linked to the page and that is commented on); and (2) *response videos*, that are multimodal comments in that they make use of different semiotic resources to respond to the master-videotext.

A study on verbal comments was carried out using a purposely created corpus, CC, and collecting comments embracing a 26-month time-span for a cross-temporal study. Written comments were chosen for analysis to allow a comparison with the results in Chapter 3, which made use of similar methods of enquiry, grounded on corpus-based lexical analysis. Spoken/written variation was explored via lexical observations, for example via keyness analysis. Keywords were studied in comparison to a general reference corpus, and its two spoken and written sub-components, to check similarities and differences with CC. Other analyses were carried out, singling out the most frequent lexical bundles, which unearthed how lexical items accrue, also shedding light on the *aboutness* of CC in general.

Contrary to the findings in the study of videochats, mainly based on qualitative observations, and in the study of community blogging, YouTube comments display lexical patterns that are very much linked to what we have defined the *master-text*, that is the foregrounded video. In the case of the video in question, "Charlie bit me," comments have been found to be highly repetitive and corpus analyses have made possible their functional categorization into six main types of groupings. Some may be also applied to YouTube *tout court*, as is the case with "Self-promotion," "Deviation" and "Removal," which can be easily found in other YouTube video comments (i.e., meta-texts), as they encompass general textual phenomena, such as self-promotional spamming or text removal. However, some specific kinds of groupings can be detected only in similar kinds of videos. For example, the "Questioning popularity" grouping is connected to the video's popularity and may not be valid for other, less popular videos. One of the most frequent and interesting groupings is represented by what we have defined as "Quotation" that has also been associated, in very broad terms, to a particular instantiation of asynchronous mode-switching. As discussed in several occasions in this book, one of the defining features of mode-switching is *synchronicity*. It is synchronicity that makes the experience rather new and unprecedented in human interaction, as the switch from oral to written discourse in asynchronous modality is very common and very well-known in a wealth of communicative events. However, the fact that some tubers like to turn video utterances into written chunks of language is rather interesting and calls for further research.

Furthermore, the principle of the *multimodal relevance maxim* has been invoked to explain the presumed low level of relevance in comment threads, interpreted as digital conversations in previous research literature (Adami 2009a, 2009b). The principle of multimodal relevance assumes that relevance needs to encompass *all* semiotic resources that are deployed in web-based interactions, even though it may also be taken for granted that a multimodal relevance maxim can also be easily found in face-to-face contexts, for example in an exchange where verbal communication is dropped in favour of a non-verbal meaning-making event, for example a gesture or a meaningful gaze. This brings us back to what occasioned the first research questions addressed in this book: how is communication changing in the light of the global digital scenario?

The self is not a monolithic and stable entity, as it is constantly fluctuating across different representations. The latter are grounded on the semiotic weight which the represented and representing subject (often but not always coinciding) decides to attribute to each *mise-en-scène*. In other words, a video may narrate the self via either a narration/confession regarding one's own personal life (e.g., by a vlog) or via segments or fragments which are considered representative of intersubjective spheres. The latter is less obviously self-narrative, as it selects subtler narrative strategies, or politics of authenticity (Wesch 2009b), if compared to a direct and first-person self-narration. However, as has been discussed in this chapter, more indirect forms of narration are overwhelmingly superior in number on YouTube. They can be analysed using a wealth of strategies in a multidisciplinary perspective. Videos such as those featuring Charlie, *MadV* or *guitar90* are *mise-en-scènes* of portions of the self, which they consider especially important for definitions of identity, respectively being a father, an illusionist and a guitarist in the cited examples.

Wesch predicts context collapse, creating special conditions for the reception of these *mise-en-scènes*, because contextual or deictic elements lose ground and much more, and at a faster rate, than in other, conventional mainstream media. Context collapses creating a textual black hole eluding well-established linguistic categories, but that strikingly exhibit patterns of syntactic repetition and lexical redundancy. Comments written by tubers are another face of representation. They could be interpreted as a desire to leave a textual trace after other people's textual vestige. The long debated and today somewhat threadbare idea of interactivity on the web takes on new semantic associations, when a multimodal text is interweaved with written or multimodal micro-textuality. Written, in the case of comments, and multimodal, in the case of response videos.

The striking number of comments referred to the discussed video show the extent and range of the phenomenon and the necessity to develop tools of analysis appropriate to this hypertrophic hyper-intertextuality. Our corpus of comments has served the purpose of allowing a foray into these metatextual phenomena. In the three weeks after the general CC was built, the video was viewed more than 5 million times, receiving further 10 thousand comments. Five millions, like all babies born in 2009 in Europe, like people living in Sicily and inhabitants of cities such as Toronto or Caracas. Less than a month later, the scenario had already changed. Every time this chapter is read, figures and statistics, views and comments will modify this arena of discussion, making every reflection ultimately unstable and infinitely evanescent.

Conclusions

This book has explored patterns of interaction in digital environments, focusing specifically on spoken and written discourse. Language mode variation has been analysed following different methods, selected according to linguistic and semiotic characteristics deployed by such multifaceted web-based texts.

The use of a range of resources has been explored using different methods, such as conversation analysis and corpus linguistics, within the umbrella framework provided by multimodal studies. The general questions addressed in this book have been cautiously answered, with the proviso that the extremely fluid nature of the web only allows tentative answers. They partially subvert possible theoretical assumptions, for example with regard to the supposed high degree of integration of semiotic resources in digital environments. The general homogeneous use of resources within specific texts has also been challenged. Rather unexpectedly, digital field work has shown that resources are less integrated than they were thought to be, and that predictions are easily contested by idiosyncratic textual practices within communities. Furthermore, predictable patterns of interaction are not easy to grasp and define, given the pace of change. Communities change, and systems of establishing and maintaining virtual relationships vary accordingly.

Verbal language still holds a privileged position in the world of communication; this is why an analysis into spoken and written variation has produced findings for future research in numerous fields of studies, such as digital literacy, media studies and linguistics, to name but the most obvious. For example, further research is needed in the context of accessibility to texts, in that specific communities (e.g., disabled people or children/teenagers coming from impoverished backgrounds) have special textual and technical needs that must be addressed. A thorough awareness of how these texts are designed and of how resources are used by communities of practice may also be useful in improving accessibility for a range of minority audiences, be it in terms of digital literacy or, more broadly, in terms of literacy *tout court*.

Spoken and written discourses, as has been amply discussed, are broader notions than those encompassed by *external factors*, such as those contrasting phonology to orthography, as claimed in Chapter 1 (Halliday 1985b, 1987). Conversely, they are ingrained in practices and activities that inform our lives and help develop our skills and pragmatic abilities as social beings. An underlying observation that runs throughout this book is that people are ready and willing to learn new practices, despite the pressure put on them by the rapidity of change. Users promptly adapt to new patterns of interaction, although they may not be fully aware that their verbal and non-verbal communication is different from everyday situations. After all, this is how language works: it adapts, and is adapted, to new communicative needs.

Adaptation is, thus, a recurring theme that underlies the forms of textuality presented in this study. Adaptation is, in other words, an overall umbrella notion that helps make sense of an apparently random and chaotic textuality. Texts are quickly adapted, usually with success, to new contexts. In turn, new contexts shape new communicative events. However, despite the claims of newness of many textual phenomena, frequently put forward in this book, the very idea of newness must be returned to its proper perspective. Just as an old aunt's letter on perfumed notepaper became an email or text message, as I said in the Introduction, a text message can be turned into a longer and more articulated piece of writing, thanks to the technical affordances of smart phones, thus paving the way for a return to more fully sustained forms of writing (cf. also Kress 2010).

Texts have been analysed as interactional practices and the protean character of such texts has been also tackled methodologically, thus adopting different but compatible strategies to capture their complexity. All resources used by participants have been taken into account, according to the encompassing rationale provided by multimodality, but verbal language has been—somewhat paradoxically—one of the most foregrounded resources. In particular, the spoken/written variation has been identified as far-ranging heuristics to investigate how web-based texts articulate different configurations and indexicality of meanings.

From this standpoint, a lexical approach has been applied to corroborate the idea that language is a system of *meaning potential* (Halliday 1978). Language, as such, is the encoding of a *behaviour potential* into a *meaning potential*, that is a means through which people can express what they *can do* by establishing relationships with other people and by converting all this into what they *can mean* (Halliday 1978; Halliday and Matthiessen 2004). What they *can mean* (i.e., the *semantic system*) is, in turn, encoded into what they *can say* (i.e., *lexicogrammar*). In accordance with this view, lexical items are not mere interchangeable entities that can be placed in fixed or pre-determined slots. Hence, lexical studies can reveal how spoken and written discourses are articulated and, to some extent, interweaved in different digital environments. Words have been studied both in *isolation* (i.e., keywords) and in *clusters* (i.e., lexical bundles) to shed light on how they function in context and how they coalesce in meaningful patterns, for example in groupings that may be ascribed more typically either to spoken or written discourse. Despite possible predictions, findings show that no ready-made formula is available to make sense of such texts, which are continuously shifting toward new patterns of socialization. However, socialization is not as far ranging and straightforward as one might predict. The possibilities of content sharing and virtual meeting of new people, thanks to the lowering of physical and virtual barriers and the availability of up-to-date technologies, do not automatically equate with greater chances of socialization. Is it possible to participate in circles beyond one's own reach? Technically yes, but socializing in digital environments may be as difficult as it is in face-to-face contexts for people who are outside the circles they wish to join. For example, some types of digital communities prevent non-acolytes from joining, just as it happens in face-to-face situations. Reasons for exclusion are numerous, as numerous as are the occasions in life when we might experience textual (and consequently social) marginalization. If we do not understand, if we do not have sufficient background information, if we are not able to successfully contribute, we may be left behind.

Albeit in different ways, videochats, blog entries and comments in media sharing communities all share features with both speech and writing. Indexical configurations are rather intricate, and dimensions of language mode variation may be identified taking into account other semiotic resources, as well. An implication that stems from this research is that spoken-like and written-like features may also encompass *other* resources. For example, the use of gaze and proxemic patterns are more typically found in spontaneous digital (video-based) conversations, whereas fonts and colours are more likely to be found in written exchanges, for example in blogs or status updates in social networks. This is not to say that these resources are implicitly deemed as subordinate to language; conversely, this observation reflects the need to consider a wider range of resources when interaction is concerned.

In keeping with the aim of this book, the paradigm of *mode-switching* has been developed to capture the novelty of the alternation of speech and writing within the same (i.e., synchronous) conversational event. This paradigm has been identified in several interactions, and patterns of mode-switching have been explored by adapting well-established notions from conversation analysis. Mode-switching can be used by participants when technical issues prevent them from continuing to effectively use the same mode (e.g., speech), for example when connection is too slow or bandwidth too small. However, other significant uses have been detected, directly related to conversational practices, as is the case of asking for a secondary floor of conversation or for repairing communication breakdown. Mode-switching has been furthermore distinguished in its use as a *self-initiated* and *other-initiated* conversational strategy, thus adapting traditional notions developed to describe turn-taking in spontaneous, live conversation.

Both one-to-one and multi-party video conversations are spontaneous, live, and meaning-making events. In videochats, communication develops along different interactional lines: participants use speech, writing, gaze, posture, kinetic action and staged proxemics to establish and maintain relationships. A transcriptional model has been proposed in this book to account for all the resources and strategies deployed by users and this model has been enriched with qualitative observations, thanks to a survey of users' own perceptions about their videochat use. This has also shown that participants learn easily and to good effect how to master all the resources they have at their disposal in order to communicate.

However powerful, language has on many occasions become secondary to other significant resources. In blog entries, for example, verbal language was dominant, but when other resources were deemed as more significant by specific users, for example by *photobloggers*, verbal language played a subservient role, that is, in captioning pictures and photos. In the study on blogs, the loose integration of resources is maybe the most striking finding and further research is needed also in the field of fanfiction, which is a popular, rapidly spreading new phenomenon, as recent blockbusters show (cf. the controversial blockbuster trilogy *Fifty Shades of Grey* by E. L. James, which was originally written and published online as fanfiction). Fans are consistently populating social networking and blogging communities. The increasing popularity of such platforms (cf. the *A-list* for blogs) and high degree of content-sharing affordances (cf. newsfeed, social networking websites) are countered by stricter notions of privacy and secrecy. This is also evident in large parts of contents that are restricted to the acolytes. Fanfiction is an emerging phenomenon that calls for further study. It typically presents fully sustained pieces of writing in blog entries or in specific web pages/websites. This fact may cast doubts on the supposed lack of skills when digital literacy is concerned, as it is suggested in some quarters that web inhabitants are at loss when it comes to committing themselves to traditional, "proper" writing. However, further evidence is needed to assess the impact and possible expansion of these practices of writing, which many creative writers contest on the grounds of their lack of originality and poor literary quality. If lexical variation, on one hand, seems to suggest that these pieces of writing are in effect wanting in traditional notions of literariness, on the other, the sample studied in this book is too limited to allow for more general claims on the matter. Furthermore, the spread and extent of the phenomenon calls for more fine-grained analyses to look at aspects such as the use of sustained prose. This contradicts assumptions with regard to the presumed inability of producing longer pieces of writing on the part of amateur writers. Furthermore, studies into lexical bundles in the corpus of blog entries have shown that language mode variation is too complex to be fully mapped.

Lexical bundles can be mainly associated with typical spoken bundles, but important deviations, such of those bundles that are mainly found in prose, point to phenomena of hybridization, and overlapping between spoken and written discourse.

The last part of the book has been devoted to the best-known video sharing community, YouTube. Applying the methods used in the chapter on blogs, commenting activities have been explored, mainly focusing on how users draw on spoken and written resources of language to establish an on-going dialogue about videos. A viral video has been sampled as a case study, "Charlie bit me," to monitor the extent of interaction among participants and to delve into language mode variation. Findings suggest that interaction among participants is very loose, whereas interaction between the single user and the video is very high. This is why the notion of *multimodal relevance maxim* has been invoked to explain the low degree of relevance among comments and the high degree of relevance between each comment and the video to which they refer. Lexical variation is relatively low, and keyness analyses have allowed functional groupings of comments, mainly based on recurring lexical keywords that have been analysed in context through concordancing. Such groupings are related to the video in question, "Charlie bit me," but they can also be applied to other similar viral videos and adjusted to explain how comments are made on *all* videos on YouTube. Studies into lexical bundles partially confirm findings from the previous chapter and examples of semantic preference and semantic prosody help complement and enrich research into language mode variation.

Patterns of communication in web-based texts are shaped by different practices, expectations, technicalities and users' skills, degree of literacy and personal passions. Language mode variation has been mainly studied via the paradigm of *mode-switching*, which has proved a useful descriptive label to encompass all the alternations between speech and writing that occur in digital interaction. However, when other resources are concerned, the notion of *resource-switching* has been used to explain different kinds of alternations, typically in *asynchronous mode*.

As explained above, the alternation of speech and writing are taking on new forms and new patterns made possible by technology, which have never before been achieved in the history of human communication. Not surprisingly, technological advances create new opportunities for interaction and communication, but a general increase in the potential for interaction (i.e., a wealth of web texts, such as blogs, chats, forums, videochats, etc.) does not necessarily equate with a proportional increase in communication. In other words, we need to be cautious in making assumptions about the extent, degree and consequences of the development of new forms of socialization. What we are concerned with here is an analysis of new ways of arranging personal and interpersonal resources, where the medium is likely to influence, if not determine, the message.

These considerations, briefly touched on here, raise questions for various disciplines, such as linguistics, semiotics, social sciences, psychology and medicine, to name but a few. Seen from another perspective, these questions point to problems that need to be addressed with the help of different scientific tools, drawn from various disciplines, further encouraging the blurring of the epistemological boundaries between them. Communication is not exclusively the domain of traditional fields involved in communication, such as linguistics. At the Interdisciplinary Conference on Linguistics in Belfast in 2011, Michael Halliday addressed the question of why we need to understand about language. He claimed that we, as human beings, inhabit two realms, that is, the realm of matter and the realm of meaning. Systems of matter, that is, *material systems*, have been theorised by the natural sciences, whereas systems of meaning, that is, *semiotic systems*, have been ignored by mainstream scientific thinking. For scientists, meaning is produced, received and gauged as *information*, but *meaning* is different from information. Meaning cannot be univocally measured and gauged. In his view, language is the most complex semiotic system and is the site of our knowing and our learning: "It is the only system complex enough to unify all the domains and divisions of human knowledge" (Halliday 2011). It is under this very broad, or maybe all-comprehensive, agenda that this book should be read. Theorising how complex semiotic systems work may reshape the way in which we envision relationships between material and semiotic systems. Shifting the balance between them may be a major breakthrough in scientific thinking.

In conclusion, mapping the galaxy of web texts, such as video conversations, blogging communities and social networking practices, is a step toward redefining patterns of interactional forms and toward identifying their possible profound implications on communication, or even on the human brain and body in the medium and in the long run. Some considerations, such as the unprecedented nature of some interactional patterns, allow us to posit that the changes we are witnessing are secondary in their impact and significance only to the invention of the technology of writing.

References

Adami, Elisabetta. 2009a. "'We/YouTube:' Exploring Sign-Making in Video-Interaction." *Visual Communication* 8 (4): 379–99.

Adami, Elisabetta. 2009b. "Video-Interaction on YouTube: Contemporary Changes in Semiosis and Communication." Paper presented at the International Symposium on Multimodal Approaches to Communication, Verona, May 2009.

Adamic, Lada A., and Natalie Glance. 2005. "The Political Blogosphere and the 2004 U.S. Election: Divided They Blog." Paper presented at the Annual Workshop on the Weblogging Ecosystem, WWW2005, Japan 2005.

Adamic, Lada A., and Eytan Adar. 2001. "You Are What You Link." Paper presented at the 10th International World Wide Web Conference, Hong Kong, China, May 2001.

Akinnaso, F. Niyi. 1985. "On the Similarities Between Spoken and Written Language." *Language and Speech* 28(4): 324–59.

Altenberg, Bengt. 1984. "Causal Linking in Spoken and Written English." *Studia Linguistica* 38: 20–69.

Althusser, Louis. 1971. "Ideology and Ideological State Apparatuses." In *Lenin and Philosophy and Other Essays*, edited by Louis Althusser, 85–131. New York: Monthly Review Press.

Andrejevic, Mark. 2009. "Exploiting Youtube: Contradictions of User-Generated Labor." In *The YouTube Reader*, edited by Pelle Snickars and Patrick Vonderau, 406–23. Stockholm: The National Library of Sweden.

Anissimov, Michael. 2012. "What Is Vlogging?" *Wisegeek.* Accessed February 3, 2012. http://www.wisegeek.com/what-is-vlogging.htm

Argyle, Michael, Roger Ingham, Florisse Alkema, and Margaret McCallin. 1973. "The Different Functions of Gaze." *Semiotica* 7(1): 19–32.

Armstrong, Cory L., and Melinda J. McAdams. 2009. "Blogs of Information: How Gender Cues and Individual Motivations Influence Perceptions of Credibility." *Journal of Computer-Mediated Communication* 14(3): 435–56.

Auty, Caroline. 2005. "UK Elected Representatives and Their Weblogs: First Impressions." *Aslib Proceedings: New Information Perspectives* 57(4): 338–55.

Bader, Jennifer. 2002. "Schriftlichkeit und Mündlichkeit in der Chat-Kommunikation." *Networx. Online Schriftenreihe des Projekts.* Accessed January 10, 2009. Sprache@web.21(29), www.mediensprache.net/de/networx/docs/networx-29.asp.

Bahnisch, Mark. 2006. "The Political Uses of Blogs." In *Uses of Blogs*, edited by Axel Bruns and Joanne Jacobs, 139–49. New York: Peter Lang.

Baldry, Anthony P., ed. 2000. *Multimodality and Multimediality in the Distance Learning Age.* Campobasso: Palladino.

Baldry, Anthony P. 2004. "Phase and Transition, Type and Instance: Patterns in Media Texts as Seen Through a Multimodal Concordancer." In *Multimodal Discourse Analysis*, edited by Kay L. O' Halloran, 83–108. London and New York: Continuum.

Baldry, Anthony P., and Paul J. Thibault. 2001. "Towards Multimodal Corpora." In *Corpora in the Description and Teaching of English*, edited by Guy Aston and Lou Burnard, 87–102. Bologna: CLUEB.

Baldry, Anthony P., and Paul J. Thibault. 2006. *Multimodal Transcription and Text Analysis. A Multimedia Toolkit with Associated On-Line Course*. London: Equinox.

Banks, Marcus. 2001. *Visual Methods in Social Research*. London: Sage.

Baron, Naomi S. 2000. *Alphabet to E-mail: How Written English Evolved and Where It's Heading*. London and New York: Routledge.

Baron, Naomi S. 2010. *Always On: Language in an Online and Mobile World*. Oxford: Oxford University Press.

Baron, Naomi S. 2013. "Reading in Print or Onscreen: Better, Worse, or About the Same?" In *Georgetown University Round Table on Languages and Linguistics 2011. Discourse 2.0: Language and New Media*, edited by Deborah Tannen and Anna Marie Trester, 201-24. Washington, DC: Georgetown University Press.

Baudrillard, Jean. (1981) 1994. *Simulacra and Simulation*. Translated by Sheila F. Glaser. Ann Arbor: University of Michigan Press.

Beaman, Karen. 1984. "Coordination and Subordination Revisited: Syntactic Complexity in Spoken and Written Narrative Discourse." In *Coherence in Spoken and Written Discourse*, edited by Deborah Tannen, 45–80. Norwood, NJ: Ablex.

Belica, Cyril. 1996. "Analysis of Temporal Change in Corpora." *International Journal of Corpus Linguistics* 1(1): 61–74.

Benjamin, Walter. (1935) 2008. *The Work of Art in the Age of Mechanical Reproduction*. Translated by J. A. Underwood. Harmonsdworth: Penguin.

Bernal, Martin. 1987. *Black Athena: The Afroasiatic Roots of Classical Civilization*. New Brunswick: Rutgers University Press.

Biber, Douglas. 1988. *Variation Across Speech and Writing*. Cambridge: Cambridge University Press.

Biber, Douglas. 1992. "On the Complexity of Discourse Complexity: A Multidimensional Analysis." *Discourse Processes* 15: 133–63.

Biber, Douglas. 1993. "Representativeness in Corpus Design." *Literary and Linguistic Computing* 8(4): 243–57.

Biber, Douglas. 1995. *Dimensions of Register Variation: A Cross-Linguistic Comparison*. New York: Cambridge University Press.

Biber, Douglas. 2003. "Lexical Bundles in Academic Speech and Writing." In *Practical Applications in Language and Computers*, edited by Barbara Lewandowska-Tomaszczyk, 165–78. Frankfurt: Peter Lang.

Biber, Douglas, and Susan Conrad. 1999. "Lexical Bundles in Conversation and Academic Prose." In *Out of Corpora: Studies in Honour of Stig Johansson*, edited by Hilde Hasselgård and Signe Oksefjell, 181–89. Amsterdam: Rodopi.

Biber, Douglas, Stig Johansson, Geoffrey Leech, Susan Conrad, and Edward Finegan. 1999. *Longman Grammar of Spoken and Written English*. London: Longman.

Biber, Douglas, Susan Conrad, and Viviana Cortés. 2004. "*If you look at...*: Lexical Bundles in University Teaching and Textbooks." *Applied Linguistics* 25(3): 371–405.

Biber, Douglas, and Barbieri, Federica. 2007. "Lexical Bundles in University Spoken and Written Registers." *English for Specific Purposes* 26: 263–86.

Blankenship, Jane. 1962. "A Linguistic Analysis of Oral and Written Style." *Quarterly Journal of Speech* 48: 419–22.

Blankenship, Jane. 1974. "The Influence of Mode, Sub-Mode, and Speaker Predilection on Style." *Speech Monographs* 41(2): 85–118.

Blood, Rebecca. 2002a. *The Weblog Handbook: Practical Advice on Creating and Maintaining your Blog*. Cambridge, MA: Perseus Publishing.

Blood, Rebecca. 2002b. "Introduction." In *We've Got Blog: How Weblogs are Changing Our Culture*, edited by John Rodzvilla, ix-xiii. Cambridge, MA: Perseus Publishing.

Blood, Rebecca. 2003. "Weblogs and Journalism: Do They Connect?" *Nieman Reports* 57(3): 61–2. Accessed February 20, 2011. http://www.nieman.harvard.edu/reports/03–3NRfall/V57N3.pdf

Bondi, Marina, and Mike Scott, eds. 2010. *Keyness in Texts*. Amsterdam and Philadelphia: John Benjamins.

Boomer, Donald S., and Allen T. Dittmann. 1964. "Speech Rate, Filled Pause and Body Movement in Interviews." *Journal of Nervous and Mental Disease* 139: 324–27.

Brown, Penelope, and Stephen C. Levinson. 1987. *Politeness: Some Universals in Language Usage*. Cambridge: Cambridge University Press.

Bruns, Axel, and Joanne Jacobs. 2006. "Introduction." In *Uses of Blogs*, edited by Axel Bruns and Joanne Jacobs, 1–8. New York: Peter Lang.

Bruns, Axel. 2006. "The Practice of News Blogging." In *Uses of Blogs*, edited by Axel Bruns and Joanne Jacobs, 11–22. New York: Peter Lang.

Cambria, Mariavita. 2011. "Opinions on the Move: An Exploration of the 'Article-cum-comments' Genre." In *Genre(s) on the Move. Hybridization and Discourse Change in Specialized Communication*, edited by Srikant Sarangi, Vanda Polese, and Giuditta Caliendo, 135-50. Napoli: Edizioni Scientifiche Italiane.

Carlson, Matt. 2007. "Blogs and Journalistic Authority." *Journalism Studies* 8: 264–79.

Carrington, Victoria, and Muriel Robinson. 2009. *Digital Literacies: Social Learning and Classroom Practices*. London: Sage.

Cashmore, Pete. 2006. "YouTube is World's Fastest Growing Website." *Mashable Business*. Accessed July 15, 2012. http://mashable.com/2006/07/22/youtube-is-worlds-fastest-growing-website/.

Cazden, Courtney B. 1989. "The Myth of Autonomous Text." In *Thinking Across Cultures: The Third International Conference on Thinking*, edited by Donald M. Topping, Doris C. Crowell, and Victor N. Kobayashi, 109–22. Hillsdale, NJ: Lawrence Erlbaum.

Chafe, Wallace C. 1982. "Integration and Involvement in Speaking, Writing, and Oral Literature." In *Spoken and Written Language: Exploring Orality and Literacy*. Vol. IX in the Series Advances in Discourse Processes, edited by Deborah Tannen, 35–53. Norwood, NJ: Ablex.

Chafe, Wallace. 1983. "Integration and Involvement in Spoken and Written Language." In *Semiotics Unfolding*, edited by Tasso Borbé, Vol. 2, 1095–102. Berlin: Mouton de Gruyter.

Chafe, Wallace. 1985. "Linguistic Differences Produced by Differences between Speaking and Writing." In *Literacy, Language and Learning*, edited by David R. Olson, Nancy Torrance, and Angela Hildyard, 105–23. Cambridge: Cambridge University Press.

Chafe, Wallace. 1988. "Punctuation and the Prosody of Written Language." *Written Communication* 5: 396–426.

Chafe, Wallace, and Jane Danielewicz. 1987. "Properties of Spoken and Written Language." In *Comprehending Oral and Written Language*, edited by Rosalind Horowitz and S. Jay Samuels, 83–113. San Diego, CA: Academic Press.

Charman, Suw. 2006. "Blogs in Business: Using Blogs Behind the Firewall." In *Uses of Blogs*, edited by Axel Bruns and Joanne Jacobs, 57–68. New York: Peter Lang.

Chien Pan, Yu, Liangwen Kuo, and Jim Jiunde Lee. 2007. "Sociability Design Guidelines for the Online Gaming Community: Role-Play and Reciprocity." In *Online Communities and Social Computing, Proceedings Second International Conference OCSE*, edited by Douglas Schuler, 426–34. London: Springer.

Chomsky, Noam. 1957. *Syntactic Structures*. Berlin: Mouton de Gruyter.

Chozick, Amy. 2012. "Two Pulitzers for Times; Huffington Post and Politico Win." *The New York Times*. April 16, 2012.

Cloud, John. 2006. "The Gurus of YouTube." *Time*. December 25, 2006. Accessed December 14, 2010. http://www.time.com/time/printout/0,8816,1570721,00. html

Cornbleet, Sandra, and Ronald Carter. 2001. *The Language of Speech and Writing*. London and New York: Routledge.

Cortés, Viviana. 2006. "Teaching Lexical Bundles in the Disciplines: An Example from a Writing Intensive History Class." *Linguistics and Education* 17: 391–406.

Coulmas, Florian. 1989. *The Writing Systems of the World*. Oxford: Basil Blackwell.

Crowley, Tony. (1989) 2003. *Standard English and the Politics of Language*. 2nd Edition. Basingstoke, Hampshire and New York: Palgrave Macmillan.

Crystal, David. 2003. *The Cambridge Encyclopedia of the English Language*. 2nd Edition. Cambridge: Cambridge University Press.

Danet, Brenda. 2010. "Computer-Mediated English." In *The Routledge Companion to English Language Studies*, edited by Janet Maybin and Joan Swann, 146–56. London and New York: Routledge.

De Kerckhove, Derrick. 1997. *The Skin of Culture. Investigating the New Electronic Reality*. London: Kogan Page.

Delwiche, Aaron. 2004. "Agenda-Setting, Opinion Leadership, and the World of Web Logs." Paper presented at the Annual Conference of the International Communication Association, New Orleans, LA, May 2004.

Denny, J. Peter. 1991. "Rational Thought in Oral Culture and Literate Decontextualization." In *Literacy and Orality*, edited by David. R. Olson and Nancy Torrance, 66–89. Cambridge: Cambridge University Press.

Derrida, Jacques. 1967. *L'écriture et la différence*. Paris: Editions du Seuil.

Derrida, Jacques. (1967) 1972. *La voix et le phénomène*. Paris: Presses Universitaires de France.

Derrida, Jacques. 1998. *Of Grammatology*. Translated by Gayatri C. Spivak. Baltimore and London: The Johns Hopkins University Press.

Devito, Joseph A. 1967a. "Levels of Abstraction in Spoken and Written Language." *Journal of Communication* 17: 354–61.

Devito, Joseph. A. 1967b. "A Linguistic Analysis of Spoken and Written Language." *Central States Speech Journal* 18: 81–5.

Drieman, Gerard H. J. 1962. "Differences between Written and Spoken Languages: An Exploratory Study." *Acta Psychologica* 20: 36–57, 78–100.

Eggins, Suzanne, and Diana Slade. (1997) 2004. *Analysing Casual Conversation*. London: Equinox.

Ekman, Paul, and Wallace V. Friesen. 1969. "The Repertoire of Nonverbal Behavior: Categories, Origins, Usage, and Coding." *Semiotica* 1: 49–98.

Ellsworth, Phoebe C., and Linda M. Ludwig. 1972. "Visual Behavior in Social Interaction." *Journal of Communication* 22(4): 375–403.

Emmison, Michael, and Philip Smith. 2000. *Researching the Visual. Images, Objects, Contexts and Interactions in Social and Cultural Inquiry*. London: Sage.

Esmaili, Kyumars Sheykh, Mohsen Jamali, Mahmood Neshati, Hassan Abolhassani, and Yasaman Soltan-Zadeh. 2006. "Experiments on Persian Weblogs." Paper presented at the Workshop Weblogging Ecosystem: Aggregation, Analysis and

Dynamics. Edinburgh, May 2006. Accessed March 23, 2012. http://www.ra.ethz. ch/CDstore/www2006/www.blogpulse.com/www2006-orkshop/papers/persian-weblogs.pdf

Etzioni, Amitai. 1993. *The Spirit of Community: The Reinvention of American Society*. New York: Touchstone.

Everts, Elisa. 2004. "Modalities of Turn-Taking in Blind/Sighted Interaction: Better to be Seen and Not Heard?" In *Discourse and Technology: Multimodal Discourse Analysis*, edited by Philip LeVine and Ronald Scollon, 128–45. Washington, DC: Georgetown University Press.

Feder, Jens. 1988. *Fractals*. New York and London: Plenum Press.

Finnegan, Ruth. (1978) 1992. *Oral Poetry: Its Nature, Significance and Social Context*. Bloomington and Indianapolis: Indiana University Press.

Firth, John Rupert. 1935. "The Technique of Semantics." *Transactions of the Philological Society*: 36–72.

Firth, John Rupert. 1951/1957. *Papers in Linguistics*. Oxford: Oxford University Press.

Flewitt, Rosie, Regine Hampel, Mirjam Hauck, and Lesley Lancaster. 2009. "What Are Multimodal Data and Transcription?" In *The Routledge Handbook of Multimodal Analysis*, edited by Carey Jewitt, 40–53. London and New York: Routledge.

Gal, Susan. 2002. "A Semiotics of the Public/Private Distinction." *Differences: A Journal of Feminist Cultural Studies* 13(1): 77–95.

Gee, James P., and Elizabeth R. Hayes. 2011. *Language and Learning in the Digital Age*. London and New York: Routledge.

Genette, Gerard. (1982) 1997. *Palimpsets. Literature at the Second Degree*. Translated by Channa Newman and Claude Doubinsky. Lincoln, NB: University of Nebraska Press.

Gibson, James Jerome. (1979) 1986. *The Ecological Approach to Visual Perception*. Hillsdale, NJ and London: Lawrence Erlbaum.

Gillmor, Dan. 2003. "Moving Toward Participatory Journalism." *Nieman Reports* 57(3): 79–80. Accessed April 4, 2012. http://www.nieman.harvard.edu/reports/03-3NRfall/V57N3.pdf

Goddard, Angela. 2004. "'The Way to Write a Phone Call:' Multimodality in Novices' Use and Perceptions on Interactive Written Discourse (IWD)." In *Discourse and Technology. Multimodal Discourse Analysis*, edited by Philip LeVine and Ron Scollon, 34–46. Washington, DC: Georgetown University Press.

Goffman, Erving. 1971. *Relations in Public: Microstudies of the Public Order*. London: Allen Lane, Penguin.

Goffman, Erving. 1981. *Forms of Talk*. Oxford: Blackwell.

Goodwin, Charles. 1980. "Restart, Pauses, and the Achievement of Mutual Gaze at Turn-Beginning." *Sociological Inquiry* 50(3–4): 272–302.

Goodwin, Charles. 1981. *Conversational Organization: Interaction Between Speakers and Hearers*. New York: Academic Press.

Goodwin, Charles. 1994. "Professional Vision." *American Anthropologist* 96(3): 606–33.

Goody, Jack. 2000. *The Power of the Written Tradition*. Washington: Smithsonian Institution Press.

Goofusgallant. 2012. "How Authors Feel about Fanfiction." *The Live Journal*. April, 19. Accessed April 19, 2012. http://ohnotheydidnt.livejournal.com/68332629. html?page=20andcut_expand=1#cutid1

Grice, H. Paul. 1952. "Logic and Conversation." In *Syntax and Semantics, 3: Speech Acts*, edited by Peter Cole and Jerry Morgan. 41–58. New York, Academic Press.

Grice, H. Paul. 1989. *Studies in the Way of Words*. Cambridge, MA: Harvard University Press.

Gross, Ralph, and Alessandro Acquisti. 2005. "Information Revelation and Privacy in Online Social Networks." Proceedings of WPES'05, 71–80. Alexandria, VA: ACM.

Grusec, Joan E., and Paul D. Hastings, eds. 2007. *Handbook of Socialization: Theory and Research*. New York: Guilford Press.

Guattari, Félix. 1989. *The Three Ecologies*. Translated by Ian Pindar and Paul Sutton. London and New York: Continuum.

Gumpertz, John J. 1982. *Discourse Strategies*. Cambridge: Cambridge University Press.

Haas, T. 2005. "From 'Public Journalism' to the 'Public's Journalism'? Rhetoric and Reality in the Discourse on Weblogs." *Journalism Studies* 6(3): 387–96.

Hall, Edward T. 1966. *The Hidden Dimension*. New York: Doubleday.

Hall, Edward T., and Mildred Reed Hall. 1990. *Understanding Cultural Differences*. Yarmouth, ME: Intercultural Press.

Halliday, Michael A.K. 1961. "Categories of the Theory of Grammar." *Word* 17: 241–92.

Halliday, Michael A.K. 1978. *Language as Social Semiotic. The Social Interpretation of Language and Meaning*. London: Edward Arnold.

Halliday, Michael A.K. 1984a. "Language as Code and Language as Behaviour: A Systemic-Functional Interpretation of the Nature and Ontogenesis of Dialogue." In *The Semiotics of Culture and Language*, edited by Robin P. Fawcett, Michael A.K. Halliday, Sydney M. Lamb, and Adam Makkai, 3–35. London: Frances Pinter. Reprinted in *The Language of Early Childhood. Collected Works of M.A.K. Halliday*, edited by Jonathan Webster, Vol. IV(2004): 227–50. London and New York: Continuum.

Halliday, Michael A.K. 1984b. "Grammatical Metaphor in English and Chinese." In *New Papers in Chinese Language Use*, edited by Beverly Hong, 9–18. Canberra: Contemporary China Centre, Australian National University (Contemporary China Papers 18). Reprinted in *Studies in Chinese Language. Collected Works of M.A.K. Halliday*, edited by Jonathan Webster, Vol. VIII(2006): 325–31. London and New York: Continuum.

Halliday, Michael A.K. 1985a. *An Introduction to Functional Grammar*. London: Edward Arnold.

Halliday, Michael A.K. 1985b. *Spoken and Written Language*. Geelong, Vic.: Deakin University Press (Language and Learning).

Halliday, Michael A.K. 1987. "Spoken and Written Modes of Meaning." In *Comprehending Oral and Written Language*, edited by Rosalind Horowitz and S. Jay Samuels, 55–82. Orlando, FL: Academic Press. Reprinted in *On Grammar. Collected Works of M.A.K. Halliday*, edited by Jonathan Webster. Vol. I(2002): 323–51. London and New York: Continuum.

Halliday, Michael A.K. 1994. *An Introduction to Functional Grammar*. 2nd Edition. London: Edward Arnold.

Halliday, Michael A.K. 1998. "Things and Relations: Regrammaticizing Experience as Technical Knowledge." In *Reading Science: Critical and Functional Perspectives on Discourse of Science*, edited by J.R. Martin, and Robert Veel. London and New York: Routledge. Reprinted in *The Language of Science. Collected Works of M.A.K. Halliday*, edited by Jonathan Webster, Vol. V(2004): 49–101. London and New York: Continuum.

Halliday, Michael A.K. 2011. "Why We Need to Understand About Language." Paper presented at the Interdisciplinary Conference on Linguistics in Belfast, October 14–15, 2011.

Halliday, Michael A.K., and Christian Matthiessen. 2004. *An Introduction to Functional Grammar*. 3rd Edition. London: Edward Arnold.

Halliday, Michael A.K., and Ruqaiya Hasan. 1976. *Cohesion in English*. London: Longman.

Harrigan, Jinni A. 2005. "Proxemics, Kinesics and Gaze." In *The New Handbook of Methods in Nonverbal Behavior Research*, edited by Jinni A. Harrigan, Robert Rosenthal, and Klaus R. Scherer, 137–98. Oxford: Oxford University Press.

Harris, Roy. 2000. *Rethinking Writing*. London and New York: Continuum.

Havelock, Eric A. 1963. *Preface to Plato*. Cambridge, MA: Harvard University Press.

Havelock, Eric A. 1982. *The Literate Revolution in Greece and Its Cultural Consequences*. Princeton, Guildford: Princeton University Press.

Havelock, Eric A. 1986. *The Muse Learns to Write: Reflections on Orality and Literacy from Antiquity to the Present*. New Haven and London: Yale University Press.

Hayes, Donald P. 1988. "Speaking and Writing: Distinct Patterns of Word Choice." *Journal of Memory and Language* 27: 572–85.

Heath, Christian, and Paul Luff. 2000. *Technology-in-Action*. New York and Cambridge: Cambridge University Press.

Heffernan, Virginia. 2006. "Who Is Interested in a Story When All Your Investment in the Characters within Is False." *New York Times*. September 7, 2006. Accessed March 3, 2012. http://themedium.blogs.nytimes.com/2006/09/07/who-is-interested-in-a-story-when-all-your-investment-in-the-characters-within-is-false/

Heller, Monica, ed. 1988. *Codeswitching: Anthropological and Sociolinguistic Perspectives*. Berlin: Mouton de Gruyter.

Herring, Susan C. 2004. "Computer-Mediated Discourse Analysis: An Approach to Researching Online Behavior." In *Designing for Virtual Communities in the Service of Learning*, edited by Sasha A. Barab, Rob Kling, and James H. Gray, 338–76. New York: Cambridge University Press.

Herring, Susan C. 2013. "Discourse in Web 2.0: Familiar, Reconfigured, and Emergent." In *Georgetown University Round Table on Languages and Linguistics 2011: Discourse 2.0: Language and New Media,* edited by Deborah Tannen and Anna Marie Trester, 1-25. Washington, DC: Georgetown University Press.

Herring, Susan C., Imma Kouper, Lois Ann Scheidt, and Elijah Wright. 2004a. "Women and Children Last: The Discursive Construction of Weblogs." *Into the Blogosphere: Rhetoric, Community, and Culture of Weblogs*, edited by Laura Gurak, Smiljana Antonijevic, Laurie Johnson, Clancy Ratliff, and Jessica Reyman. Accessed August 18, 2012. http://blog.lib.umn.edu/blogosphere/women_and_children.html

Herring, Susan C., Lois Ann Scheidt, Sabrina B. Bonus, and Elijah Wright. 2004b. "Bridging the Gap: A Genre Analysis of Weblogs." Paper presented at the Hawaii International Conference on System Sciences. Los Alamitos: IEEE Computer Society Press. Accessed April 12, 2012. http://ella.slis.indiana.edu/~herring/herring.scheidt.2004.pdf

Herring, Susan C., Lois Ann Scheidt, Sabrina Bonus, and Elijah Wright. 2005a. "Weblogs as a Bridging Genre." *Information, Technology and People* 18(2): 142–71.

Herring, Susan C., Inna Kouper, John C. Paolillo, Lois Ann Scheidt, Michael Tyworth, Peter Welsch, Elijah Wright, and Ning Yu. 2005b. "Conversations in the Blogosphere: An Analysis 'From the Bottom Up.'" Proceedings of the Thirty-Eighth Hawai'i International Conference on System Sciences (HICSS-38). Los Alamitos: IEEE Computer Society Press.

Herring, Susan C., and John C. Paolillo. 2006. "Gender and Genre Variation in Weblogs." *Journal of Sociolinguistics* 10(4): 439–59.

Herring, Susan C., Lois Ann Scheidt, Inna Kouper, and Elijah Wright. 2007. "A Longitudinal Content Analysis of Weblogs: 2003-2004." In *Blogging, Citizenship, and the Future of Media,* edited by Mark Tremayne, 3–20. London and New York: Routledge.

Hesch, June. I. 1982. "Predication Typing of Oral and Written Expository Genre." In *The Eight LACUS Forum 1981*, edited by Waldemar Gutwinski and Grace Jolly, 471–80. Columbia, SC: Hornbeam.

Hewings, Ann, and Martin Hewings. 2005. *Grammar and Context*. Routledge Applied Linguistics. London and New York: Routledge.

Hine, Christine. 2000. *Virtual Ethnography*. London: Sage.

Hobsbawn, Eric, and Terence Ranger, eds. 1983. *The Invention of Tradition*. Cambridge: Cambridge University Press.

Hodkinson, Paul. 2006. "Subcultural Blogging? Online Journals and Group Involvement Among U.K. Goths." In *Uses of Blogs*, edited by Axel Bruns and Joanne Jacobs. 187–98. New York: Peter Lang.

Hoey, Michael. 2004. "Lexical Priming and the Properties of Text." In *Corpora and Discourse*, edited by Alan Partington, John Morley, and Louann Haarman, 385–412. Bern: Peter Lang.

Hoey, Michael. 2005. *Lexical Priming. A New Theory of Words and Language*. London and New York: Routledge.

Hogg, Richard M., Norman F. Blake, and John Algeo. 2001. *The Cambridge History of the English Language*. Vols. 1–6. Cambridge: Cambridge University Press.

Holmes, David. 2005. *Communication Theory. Media, Technology and Society*. London: Sage.

Hutchby, Ian, and Robin Wooffitt. (1998) 2008. *Conversation Analysis*. 2nd Edition. Cambridge: Polity Press.

Jahandarie, Khosrow. 1999. *Spoken and Written Discourse: A Multi-Disciplinary Perspective*. Stamford, CT: Ablex.

Jenkins, Henry. 2006. *Fans, Bloggers, and Gamers: Exploring Participatory Culture*. New York: New York University Press.

Jewitt, Carey. 2009. *The Routledge Handbook of Multimodal Analysis*. London and New York: Routledge.

Johnson, Thomas. J., and Barbara K. Kaye. 2000. "Using is Believing: The Influence of Reliance on the Credibility of Online Political Information Among Politically Interested Internet Users." *Journalism and Mass Communication Quarterly* 77: 865–79.

Johnson, Thomas. J., Barbara K. Kaye, Shannon L. Bichard, and W. Joann Wong. 2008. "Every Blog Has Its Day: Politically-Interested Internet Users' Perceptions of Blog Credibility." *Journal of Computer-Mediated Communication* 13: 110–22.

Jones, Andrew, Magnus Lang, Graham Fyffe, Xueming Yu, Jay Busch, Ian McDowall, Mark T. Bolas, and Paul E. Debevec. 2009. "Achieving Eye Contact in a One-to-Many 3D Video Teleconferencing System." Accessed October 15, 2010. http://arnetminer.org/viewpub.do?pid=1227884.

Jones, Graham M., and Bambi B. Schieffelin. 2009. "Talking Text and Talking Back: 'My BFF Jill' from Boob Tube to YouTube." *Journal of Computer-Mediated Communication* 14: 1050–79.

Kaye, Barbara. 2005. "It's a Blog, Blog, Blog World: Users and Uses of Weblogs." *Atlantic Journal of Communication* 13(2): 73–95.

Kaye, Barbara K., Thomas J. Johnson. 2004a. "Weblogs as a Source of Information about the 2003 Iraq War." In *Global Media Go to War: Role of News and Entertainment Media During the 2003 Iraq War*, edited by Ralph D. Berenger, 291–301. Spokane, WA: Marquette Books.

Kaye, Barbara. K., and Thomas J. Johnson. 2004b. "A Web for All Reasons: Use and Gratifications of Internet Components for Political Information." *Telematics and Informatics* 21: 197–223.

Kelleher, Tom, and Barbara M. Miller. 2006. "Organizational Blogs and the Human Voice: Relational Strategies and Relational Outcomes." *Journal of Computer-Mediated Communication* 11(2), article 1. Accessed 23 March, 2012. http://jcmc.indiana.edu/vol11/issue2/kelleher.html

Kendon, Adam. 1967. "Some Functions of Gaze-Direction in Social Interaction." *Acta Psychologica* 26: 22–63.

Kendon, Adam. 1990. *Conducting Interaction. Patterns of Behavior in Focused Encounters.* Cambridge: Cambridge University Press.

Kendon, Adam. 2004. *Gesture: Visible Action as Utterance.* Cambridge: Cambridge University Press.

Kenix, Linda J. 2009. "Blogs as Alternative." *Journal of Computer-Mediated Communication* 14(4): 790–822.

Kessler, Frank, and Mirko Tobias Schäfer. 2009. "Navigating YouTube: Constituting a Hybrid Information Management System." In *The YouTube Reader*, edited by Pelle Snickars and Patrick Vonderau, 275–91. Stockholm: The National Library of Sweden.

Knox, John S. 2007. "Visual-verbal Communication on Online Newspaper Home Pages." *Visual Communication* 6(1): 19–53.

Knox, John S. 2009. "Punctuating the Home Page: Image as Language in an Online Newspaper." *Discourse and Communication* 3(2): 145–72.

Knox, John S. 2009. "Visual Minimalism in Hard News: Thumbnail Faces on the smh Online Home Page." *Social Semiotics* 19(2): 165–89.

Kopytoff, Verne G. 2011. "Blogs Wane as the Young Drift to Sites Like Twitter." *The New York Times.* February 20, 2011. Accessed March 20, 2012. http://www.nytimes.com/2011/02/21/technology/internet/21blog.html?_r=1

Kress, Gunther 2003. *Literacy in the New Media Age.* London and New York: Routledge.

Kress, Gunther. 2010. *Multimodality. A Social Semiotic Approach to Contemporary Communication.* London and New York: Routledge.

Kress, Gunther, and Theo van Leeuwen. 1996. *Reading Images. The Grammar of Visual Design.* London and New York: Routledge.

Kress, Gunther, and Theo van Leeuwen. 2001. *Multimodal Discourse. The Modes and Media of Contemporary Communication.* London: Edward Arnold.

Kress, Gunther, and Theo van Leeuwen. 2002. "Colour as a Semiotic Mode: Notes for a Grammar of Colour." *Visual Communication* 1(3): 343–68.

Kress, Gunther, and Theo van Leeuwen. 2006. *Reading Images. The Grammar of Visual Design.* 2nd edition. London and New York: Routledge.

Kress, Gunther, Carey Jewitt, Jon Ogborn, and Charalampos Tsatsarelis. 2001. *Multimodal Teaching and Learning: The Rhetorics of the Science Classroom.* London and New York: Continuum.

Kress, Gunther, Carey Jewitt, Jill Bourne, Anton Franks, John Hardcastle, Ken Jones, and Euan Reid. 2004. *English in Urban Classrooms. A Multimodal Perspective on Learning and Teaching.* London and New York: Routledge.

Lakoff, Robin Tolmach. 1982. "Some of My Favorite Writers Are Literate: The Mingling of Oral and Literate Strategies in Written Communication." In *Spoken and Written Language: Exploring Orality and Literacy.* Vol. IX in the Series Advances in Discourse Processes, edited by Deborah Tannen, 239–60. Norwood, NJ: Ablex.

Lange, Patricia G. 2008. "Publicly Private and Privately Public: Social Networking on YouTube." *Journal of Computer-Mediated Communication* 13: 361–80.

Lange, Patricia G. 2009. "Videos of Affinity on YouTube." In *The YouTube Reader*, edited by Pelle Snickars and Patrick Vonderau, 70–88. Stockholm: The National Library of Sweden.

Lanier, Jaron. 2006. *Information Is an Alienated Experience.* New York: Basic Books.

Lanier, Jaron. 2010. *You Are Not a Gadget: A Manifesto.* New York: Alfred A. Knopf.

Lankshear, Colin, and Michele Knobel. 2006. *New Literacies: Everyday Practices and Classroom Learning.* Philadelphia: Open University Press.

Lasica, Joseph D. 2002a. "Blogging as a Form of Journalism." *USC Annenberg Online Journalism Review.* Accessed February 3, 2012. http://www.ojr.org/ojr/lasica/1019166956.php

Lasica, Joseph D. 2002b. "Weblogs: A New Source of News." In *We've Got Blog: How Weblogs are Changing Our Culture*, edited by Rebecca Blood, 171–82. Cambridge, MA: Perseus Publishing.

Lasica, Joseph D. 2003. "Blogs and Journalism Need Each Other." *Nieman Reports,* 57(3), 70–3. Accessed October 3, 2010. http://www.nieman.harvard.edu/reports/03–3NRfall/V57N3.pdf

Laver, John. 1975. "Communicative Functions of Phatic Communion." In *The Organisation of Behaviour in Face-to-Face Interaction*, edited by Adam Kendon, Richard Harris, and Mary R. Key, 215–38. The Hague: Mouton.

Leander, Kevin, and Lalitha Vasudevan. 2009. "Multimodality and Mobile Culture." In *The Routledge Handbook of Multimodal Analysis*, edited by Carey Jewitt, 127–39. London and New York: Routledge.

Lenhart, Amanda, and Susannah Fox. 2006. "Bloggers: A Portrait of the Internet's New Storytellers." *The Pew Internet and American Life Project.* Accessed March 22, 2012. http://www.pewinternet.org/~/media//Files/Reports/2006/PIP%20Bloggers%20Report%20July%2019%202006.pdf.

Lenhart, Amanda, Kristen Purcell, Aaron Smith, and Kathryn Zickuhr. 2010. "Social Media and Mobile Internet Use Among Teens and Young Adults." *The Pew Internet and American Life Project*. Accessed February 3, 2012. http://www.pewinternet.org/Reports/2010/Social-Media-and-Young-Adults/Summary-of-Findings.aspx

Li, Dan. 2007. "Why Do You Blog: A Uses-and-Gratifications Inquiry into Bloggers' Motivations." Paper presented at the Annual Meeting of the International Communication Association, TBA, San Francisco, CA, May 24, 2007. Accessed March 23, 2012. http://citation.allacademic.com/meta/p_mla_apa_research_citation/1/7/1/4/9/p171490_index.html

Lieberman, Philip. 1975. *On the Origins of Language: An Introduction to the Evolution of Human Speech*. New York: Macmillan.

Lofland, Lyn. 1973. *A World of Strangers*. New York: Basic Books.

Malinowski, Bronislaw. 1923. "The Problem of Meaning in Primitive Languages." In *The Meaning of Meaning*, edited by Charles K. Ogden and Ivor A. Richards, 296–336. London and New York: Routledge.

Marsh, Jackie, ed. 2005. *Popular Culture, New Media and Digital Literacy in Early Childhood*. London and New York: Routledge.

Martin, Jim R., and Peter R.R. White. 2005. *The Language of Evaluation. Appraisal in English*. Basingstoke: Palgrave Macmillan.

Martinec, Radan. 2004. "Gestures that Co-Occur with Speech as a Systematic Resource: The Realization of Experiential Meanings in Indexes." *Social Semiotics* 14 (2): 193–213.

Martinec, Radan, and Theo van Leeuwen. 2009. *The Language of New Media Design. Theory and Practice*. London and New York: Routledge.

Marx, Carl. (1867) 1976. *The Capital*. Vol. 1. London: Penguin.

McCarthy, Michael. 2006. " 'This That and the Other:' Multi-Word Clusters in Spoken English as Visible Patterns of Interaction." In Michael McCarthy. *Explorations in Corpus Linguistics*, 7–26. Cambridge: Cambridge University Press.

McDonald, Paul. 2009. "Digital Discords in the Online Media Economy: Advertising versus Content versus Copyright." In *The YouTube Reader*, edited by Pelle Snickars and Patrick Vonderau, 387–405. Stockholm: The National Library of Sweden.

McEnery, Anthony, and Andrew Wilson. 2001. *Corpus Linguistics*. 2nd Edition. Edinburgh: Edinburgh University Press.

McEnery, Anthony, Richard Xiao, and Yukio Tono. 2006. *Corpus-based Language Studies*. Routledge Applied Linguistics Series. London and New York: Routledge.

McKenna, Michael C., and Janet C. Richards, eds. 2003. *Integrating Multiple Literacies in K-8 Classrooms: Cases, Commentaries, and Practical Applications*. Hillsdale, NJ: Lawrence Erlbaum.

McLaughlin, Margaret L., Kerry K. Osborne, and Christine B. Smith. 1995. "Standards of Conduct on Usenet." In *Cybersociety: Revisiting Computer-Mediated Communication and Community*, edited by Steven Jones, 90–111. Thousand Oaks, CA: Sage.

McLuhan, Marshall. 1962. *The Gutenberg Galaxy: The Making of Typographic Man*. Toronto, Canada: University of Toronto Press.

McLuhan, Marshall. 1964. *Understanding Media: The Extensions of Man*. New York: McGraw Hill.

McLuhan, Marshall, and Quentin Fiore. 1967. *The Medium Is the Massage: An Inventory of Effects*. London: Penguin.

Melanson, G. 2012. "What Is a Vlogger?" *Wisegeek*. Accessed February 3, 2012. http://www.wisegeek.com/what-is-a-vlogger.htm

Merelo, Juan Julián, José Luis Orihuela, Victor Ruiz, and Fernando Tricas. 2004. "Revisiting the Spanish Blogosphere." In *BlogTalks 2*, edited by Thomas N. Burg, 339–52. Norderstedt: Books on Demand.

Metz, Christian. (1980) 1993. *Le signifiant imaginaire. Psychanalyse et cinéma*. Paris: Christian Bourgois Éditeur.

Metzger, Miriam J., Andrew J. Flanagin, Karen Eyal, Daisy R. Lemus, and Robert McCann. 2003. "Credibility for the 21st Century: Integrating Perspectives on Source, Message, and Media Credibility in the Contemporary Media Environment." In *Communication Yearbook 27*, edited by Pamela J. Kalbfleisch, 293–335. Hillsdale, NJ: Lawrence Erlbaum.

Miller, B. 2012. "What Does a Blogger Do?" *Wisegeek*. Accessed February 3, 2012. http://www.wisegeek.com/what-does-a-blogger-do.htm

Miller, Carolyn R., and Dawn Shepherd. 2004. "Blogging as Social Action: A Genre Analysis of the Weblog." In *Into the Blogosphere: Rhetoric, Community, and Culture of Weblogs*, edited by Laura Gurak, Smiljana Antonijevic, Laurie Johnson, Clancy Ratliff, and Jessica Reyman. University of Minnesota. Accessed February 3, 2012. http://blog.lib.umn.edu/blogosphere/blogging_as_social_action_a_genre_analysis_of_the_weblog.html

Miller, Jim. 1993. "Spoken and Written Language: Language Acquisition and Literacy." In *Literacy and Language Analysis*, edited by Robert J. Scholes, 99–141. Hillsdale, NJ: Lawrence Erlbaum.

Miller, Jim, and Regina Weinert. 1998. *Spontaneous Spoken Language: Syntax and Discourse*. Oxford: Clarendon Press.

Miura, Asako, and Kiyomi Yamashita. 2007. "Psychological and Social Influences on Blog Writing: An Online Survey of Blog Authors in Japan." *Journal of Computer-Mediated Communication* 12(4): 1452–71.

Moreno Jaén, María. 2009. Recopilación, desarrollo pedagógico y evalución de un banco de colocaciones frecuentes de la lengua inglesa a través de la lingüística de corpus y computacional. PhD dissertation. University of Granada.

Mortensen, David C., ed. 2009. *Communication Theory*. 2nd Edition. New Brunswick, NJ: Transaction Publishing.

Nardi, Bonnie A., Diane J. Schiano, and Michelle Gumbrecht. 2004. "Blogging as Social Activity, Or, Would You Let 900 Million People Read Your Diary?" *Proceedings of Computer Supported Cooperative Work 2004*. Accessed February 3, 2012. http://home.comcast.net/%7Ediane.schiano/CSCW04.Blog.pdf

Ng, Deborah. 2012. "What Are Blogs?" *Wisegeek*. Accessed February 3, 2012. http://www.wisegeek.com/what-are-blogs.htm

Nickerson, Raymond S. 1981. "Speech Understanding and Reading: Some Differences and Similarities." In *Perception of Print: Reading Research in Experimental Psychology*, edited by Ovid J. L. Tzeng and Harry Singer, 257–89. Hillsdale, NJ: Lawrence Erlbaum.

Nmincite. 2012a. "Buzz in the Blogosphere: Millions More Bloggers and Blog Readers." March 8, 2012. Accessed March 23, 2012. http://www.nmincite.com/?p=6531

Nmincite. 2012b. "State of the Media: U.S. Digital Consumer Report." Accessed March, 23 2012. http://www.nielsen.com/content/dam/corporate/us/en/reports-downloads/2012-Reports/Digital-Consumer-Report-Q4–2012.pdf

Norris, Sigrid. 2004. *Analyzing Multimodal Interaction. A Methodological Framework*. London and New York: Routledge.

Nowson, Scott, and Jon Oberlander. 2006. "The Identity of Bloggers: Openness and Gender in Personal Weblogs." Paper presented at AAAI Spring Symposium, Computational Approaches to Analysing Weblogs, Stanford University, 2006. Accessed March 23, 2012. http://aaaipress.org/Papers/Symposia/Spring/2006/SS-06–03/SS06–03–032.pdf

O'Donnell, Susan, Kerry Gibson, Mary Milliken, and Janice Singer. 2008. "Reacting to YouTube Videos: Exploring Differences Among User Groups." Proceedings of the International Communications Association Annual Conference (ICA 2008) Montreal, Quebec, May 22–26, 2008. http://iit-iti.nrc-cnrc.gc.ca/iit-publications-iti/docs/NRC-50361.pdf.

O'Halloran, Kay L., ed. 2004. *Multimodal Discourse Analysis: Systemic Functional Perspectives*. London and New York: Continuum.

O'Halloran, Kay L. 2005. *Mathematical Discourse. Language, Symbolism and Visual Images*. London and New York: Continuum.

O'Halloran, Kay L., and Bradley A. Smith, eds. 2011. *Multimodal Studies: Exploring Issues and Domains*. London and New York: Routledge.

O'Halloran, Kay L., Sabine Tan, Bradley A. Smith, and Alexey A. Podlasov. 2010. "Challenges in Designing Digital Interfaces for the Study of Multimodal Phenomena." *Information Design Journal* 18(1): 2–21.

Online Etymology Dictionary, s.v. "text." Accessed July 19, 2011. http://www.etymonline.com/index.php?term=text&allowed_in_frame=0

O'Toole, Michael. 2010. *The Language of Displayed Art*. 2nd Edition. London and New York: Routledge.

Ochs, Elinor. 1979. "Planned and Unplanned Discourse." In *Discourse and Syntax*, edited by Talmy Givón, 51–80. New York: Academic Press.

Okpewho, Isidore. 1979. *The Epic in Africa*. New York: Columbia University Press.

Okpewho, Isidore. 1992. *African Oral Literature: Backgrounds, Character and Continuity*. Bloomington and Indianapolis: Indiana University Press.

Olson, David R. 1977. "From Utterance to Text: The Bias of Language in Speech and Writing." *Harvard Educational Review* 47: 257–81.

Ong, Walter J. 1967. *The Presence of the Word*. New Haven: Yale University Press.

Ong, Walter J. 1977. *Interfaces of the Word*. Ithaca, NY: Cornell University Press.

Ong, Walter J. 1982. *Orality and Literacy: The Technologizing of the Word*. London: Methuen.

Ong, Walter J. 1992. "Writing is a Technology that Restructures Thought." In *The Linguistics of Literacy*, edited by Pamela Downing, Susan D. Lima, and Michael Noonan, 293–319. Amsterdam and Philadelphia: John Benjamins.

Otto, Maximilian, John P. Lewis, and Ingemar Cox. 1993. "Teleconferencing Eye Contact Using a Virtual Camera." Proceedings CHI '93 INTERACT '93 and CHI '93 Conference Companion on Human Factors in Computing Systems, 109–10. Amsterdam. Accessed October 15, 2010. http://www0.cs.ucl.ac.uk/staff/I.Cox/Content/papers/1993/ic93.pdf

Papacharissi, Zizi. 2004. "The Blogger Revolution? Audiences as Media Producers." Paper presented at the Annual Conference of the International Communication Association, New Orleans, May 2004.

Park, David. 2003. "Bloggers and Warbloggers as Public Intellectuals: Charging the Authoritative Space of the Weblog." Paper presented at Internet Research 4.0, Toronto, October 2003.

Partington, Alan. 1998. *Patterns and Meanings*. Amsterdam and Philadelphia: John Benjamins.

Partington, Alan. 2004. "'Utterly Content in Each Other's Company:' Semantic Prosody and Semantic Preference." *International Journal of Corpus Linguistics* 9(1): 131–56.

Pedersen, Sarah, and Caroline Macafee. 2007. "Gender Differences in British Blogging." *Journal of Computer-Mediated Communication* 12(4): 1472–92.

Piazza, Roberta, Monika Bednarek, and Fabio Rossi, eds. 2011. *Telecinematic Discourse. Approaches to the Language of Films and Television Series*. Amsterdam and Philadelphia: John Benjamins.

Pink, Sarah. (2001) 2007. *Doing Visual Ethnography*. 2nd Edition. London: Sage.

Poster, Mark. 1995. *The Second Media Age*. Cambridge: Polity Press.

Poster, Mark. 1997. "Cyberdemocracy: Internet and the Public Sphere." In *Internet Culture*, edited by David Porter. 201–17. New York and London: Routledge.

Postman, Neil. 1985. *Amusing Ourselves to Death: Public Discourse in the Age of Show Business*. New York: Penguin.

Prosser, Jon, ed. 1998. *Image-Based Research*. London and New York: Routledge.

Purcell, Patrick. A., ed. 2006. *Networked Neighborhoods. The Connected Community in Context*. London: Springer.

Qian, Hua, and Craig R. Scott. 2007. "Anonymity and Self-Disclosure on Weblogs." *Journal of Computer-Mediated Communication* 12:1428–51.

Rheingold, Howard. 1993. *The Virtual Community: Homesteading on the Electronic Frontier*. London: Secker and Warburg.

Roach, Peter. 2000. *English Phonetics and Phonology*. 3rd Edition. Cambridge: Cambridge University Press.

Rogers, Richard, and Noortje Marres. 2002. "French Scandals on the Web, and on the Streets: A Small Experiment in Stretching the Limits of Reported Reality." *Asian Journal of Social Science* 30(2): 339–53.

Rose, Gillian. 2001. *Visual Methodologies. An Introduction to the Interpretation of Visual Materials*. London: Sage.

Rosen, Jay. 2011. "Why 'Bloggers vs. Journalists' Is Still with Us." *Press Think*. Accessed April 13, 2012. http://pressthink.org/2011/03/monsters-of-the-newsroom-id-why-bloggers-vs-journalists-is-still-with-us/

Rosenberg, Scott, ed. 2010. *Say Everything: How Blogging Began, What It's Becoming and Why It Matters*. New York: Three Rivers Press.

Rouse, Roger. 1991. "Mexican Migration and the Social Space of Postmodernism." *Diaspora* 1(1): 8–24.

Rundle, Michael. 2012. "Policing Racism Online: Liam Stacey, YouTube And The Law Of Big Numbers." April 7, 2012. *The Huffington Post*. Accessed July 15, 2012. http://www.huffingtonpost.co.uk/2012/04/06/policing-racism-online-liam-stacey-fabrice-muamba-abuse-twitter-youtube-facebook_n_1407795.html

Saakana, Amon Saba. 1996. *Colonization and the Destruction of the Mind*. London: Karnak House.

Sacks, Harvey. 1992. *Lectures on Conversation*. Oxford: Blackwell.

Sacks, Harvey, Emanuel A. Schegloff, and Gail Jefferson. 1974. "A Simplest Systematics for the Organization of Turn-Taking for Conversation." *Language* 50: 696–735.

Said, Edward. (1978) 1995. *Orientalism: Western Conceptions of the Orient*. Harmondsworth: Penguin.

Said, Edward. 1993. *Culture and Imperialism*. New York: Knopf.

Sanderson, Jimmy. 2008. "The Blog is Serving Its Purpose: Self-Presentation Strategies on 38pitches.com." *Journal of Computer-Mediated Communication* 13(4): 912–36.

Schachtebeck, Thomas. 2005. "Internet Chat Communication: A Tightrope Walk Between Oral Communication and Written Communication." Seminar Paper. Grin Verlag.

Schegloff, Emanuel A., Gail Jefferson, and Harvey Sacks. 1977. "The Preference for Self-Correction in the Organization of Repair in Conversation." *Language* 53(2): 361–82.

Schiffrin, Deborah. 1987. *Discourse Markers. Studies in Interactional Sociolinguistics 5*. Cambridge: Cambridge University Press.

Schmidt, Jan. 2007. "Blogging Practices: An Analytical Framework." *Journal of Computer-Mediated Communication* 12:1409–27.

Scoble, Robert, and Shel Israel. 2006. *Naked Conversations: How Blogs are Changing the Way Businesses Talk with Customers*. Hoboken, NJ: Wiley.

Scollon, Ron. 1998. *Mediated Discourse as Social Interaction: A Study of News Discourse*. London: Longman.

Scollon, Ron. 2001. *Mediated Discourse. The Nexus of Practice*. London and New York: Routledge.

Scollon, Ron, and Philip LeVine, eds. 2004. *Discourse and Technology. Multimodal Discourse Analysis*. Washington, DC: Georgetown University Press.

Scollon, Ron, and Suzanne Wong Scollon. 1984. "Cooking It Up and Boiling It Down: Abstracts in Athabaskan Children's Story Retellings." In *Coherence in Spoken and Written Discourse*, edited by Deborah Tannen, 173–97. Norwood, NJ: Ablex.

Scollon, Ron, and Suzanne Wong Scollon. 2009. "Multimodality and Language. A Retrospective and Prospective View." In *The Routledge Handbook of Multimodal Analysis*, edited by Carey Jewitt, 170–80. London and New York: Routledge.

Scott, Mike. 2008. *WordSmith Tools version 5*. Liverpool: Lexical Analysis Software.

Scott, Mike. 2012. *WordSmith Tools version 6*. Liverpool: Lexical Analysis Software.

Scott, Mike, and Christopher Tribble. 2006. *Textual Patterns. Key Words and Corpus Analysis in Language Education*. Amsterdam and Philadelphia: John Benjamins.

Seifert, Uwe, Jin Hyun Kim, and Anthony Moore, eds. 2008. *Paradoxes of Interactivity: Perspectives for Media Theory, Human-computer Interaction, and Artistic Investigations*. Berlin: Bielefeld Verlag.

Selber, Stuart. 2004. *Multiliteracies for a Digital Age*. Carbondale: Southern Illinois University Press.

Selwyn, Neil. 2011. *Education and Technology: Key Issues and Debates*. London and New York: Continuum.

Sennett, Richard. 1977. *The Fall of Public Man*. Cambridge: Cambridge University Press.

Shettleworth, Sarah J. 1988. *Cognition, Evolution, and Behavior*. Oxford and New York: Oxford University Press.

Simmel, Georg. (1908) 1921. "Sociology of the Senses: Visual Interaction." In *Introduction to the Science of Sociology*, edited by Robert E. Park and Edward Burgess, 356–61. Chicago: Chicago University Press.

Simmel, Georg. 1949. "The Sociology of Sociability." *The American Journal of Sociology* 55: 254–61.

Sinclair, John. 1966. "Beginning the Study of Lexis." In *In Memory of J.R. Firth*, edited by Charles Bazell, John Catford, Michael Halliday, and Robert Robins, 410–30. London: Longman.

Sinclair, John. 1996. "The Search for Units of Meaning." *Textus* 9(1): 75–106.

Sinclair, John. 1998. "The Lexical Item." In *Contrastive Lexical Semantics*, edited by Edda Weigand, 1–24. Philadelphia and Amsterdam: John Benjamins.

Sinclair, John. 2003. *Reading Concordances*. London: Longman.

Sindoni, Maria Grazia. 2010. "Models of Verbal and Non-Verbal Interaction in Web 2.0 Textuality." *Bérénice* 16 (43): 85–93.

Sindoni, Maria Grazia. 2011a. "Online Conversations. A Sociolinguistic Investigation into Young Adults' Use of Videochats." *Classroom Discourse* 2(2): 219–35.

Sindoni, Maria Grazia. 2011b. *Systemic-Functional Grammar and Multimodal Studies. An Introduction with Text Analysis.* Como and Pavia: Ibis.

Sindoni, Maria Grazia. 2011c. Mise-en-scène. Politiche di auto-rappresentazione e autenticità su YouTube. *Mantichora* 1: 617–40.

Sindoni, Maria Grazia. 2012. "Mode-Switching: How Oral and Written Modes Alternate in Videochats." In *Web Genres and Web Tools with Contributions from the Living Knowledge Project*, edited by Mariavita Cambria, Cristina Arizzi, and Francesca Coccetta, 141–53. Como and Pavia: Ibis.

Smith, Larry W. 1994. "An Interactionist Approach to the Analysis of Similarities and Differences between Spoken and Written Language." In *Sociocultural Approaches to Language and Literacy*, edited by Vera John-Steiner, Carolyn P. Panofsky, and Larry W. Smith, 43–81. New York, NY: Cambridge University Press.

Smith, S. E. 2012. "What Is a Bleg?" *Wisegeek*. Accessed February 3, 2012. http://www.wisegeek.com/what-is-a-bleg.htm

Snickars, Pelle, and Patrick Vondereau, eds. 2009. *The YouTube Reader.* Stockholm: National Library of Sweden.

St. Amant, Robert, and Thomas E. Horton. 2008. "Revisiting the Definition of Animal Tool Use." Accessed January 10, 2011. ftp://ftp.ncsu.edu/pub/unity/lockers/ftp/csc_anon/tech/2007/TR-2007–16.pdf. 1–20.

Stefanone, Michael A., and Chyng-Yang Jang. 2008. "Writing for Friends and Family: The Interpersonal Nature of Blogs." *Journal of Computer-Mediated Communication* 13(1): 123–40.

Stein, Dieter, ed. 1992. *Cooperating with Written Texts. The Pragmatics and Comprehension of Written Texts.* Studies in Anthropological Linguistics. Berlin: Mouton de Gruyter.

Stelarc. 1991. "Prosthetics, Robotics, and Remote Existence: Postevolutionary Strategies." *Leonardo* 24 (5): 591–95.

Stone, Biz. 2004. *Who Let the Blogs Out? A Hyperconnected Peek at the World of Weblogs.* New York: St. Martin's Press.

Stubbs, Michael. 1996. *Text and Corpus Analysis. Computer-Assisted Studies of Language and Culture.* Oxford and Cambridge, MA: Blackwell.

Stubbs, Michael. 2001. *Words and Phrases. Corpus Studies of Lexical Semantics.* Oxford: Blackwell.

Stubbs, Michael. 2002. "Two Quantitative Methods of Studying Phraseology in English." *International Journal of Corpus Linguistics* 7(2): 215–44.

Tannen, Deborah, ed. 1982a. *Spoken and Written Language: Exploring Orality and Literacy.* Vol. IX in the Series Advances in Discourse Processes. Norwood, NJ: Ablex.

Tannen, Deborah. 1982b. "The Oral/Literate Continuum in Discourse." In *Spoken and Written Language: Exploring Orality and Literacy.* Vol. IX in the Series Advances in Discourse Processes, edited by Deborah Tannen, 1–15. Norwood, NJ: Ablex.

Tannen, Deborah. 1989. *Talking Voices. Repetition, Dialogue, and Imagery in Conversational Discourse.* Cambridge: Cambridge University Press.

Tannen, Deborah, and Anna Marie Trester, eds. 2013. *Georgetown University Round Table on Languages and Linguistics 2011. Discourse 2.0: Language and New Media.* Washington, DC: Georgetown University Press.

Tateo, Luca. 2005. "The Italian Extreme Right On-Line Network: An Exploratory Study Using an Integrated Social Network Analysis and Content Analysis Approach." *Journal of Computer-Mediated Communication* 10(2): article 10. Accessed March 23, 2012. http://jcmc.indiana.edu/vol10/issue2/tateo.html

Tattersall, Ian. 1995. *The Fossil Trail: How We Know What We Think We Know About Human Evolution.* New York: Oxford University Press.

Technorati. 2011. "State of the Blogosphere Report 2011." Accessed February 3, 2012. http://technorati.com/blogging/article/state-of-the-blogosphere-2011-introduction/

ten Have, Paul. 1999. *Doing Conversation Analysis: A Practical Guide*. London: Sage.

The Daily Telegraph. 2008. "YouTube's Worst Comments Blocked by Filter." September 2, 2008. Accessed July 14, 2012. http://www.telegraph.co.uk/news/newstopics/howaboutthat/2668997/YouTubes-worst-comments-blocked-by-filter.html#.

The Guardian. 2009. "Our Top 10 Funniest YouTube Comments—What Are Yours?" November 3. Accessed July 14, 2012. http://www.guardian.co.uk/technology/blog/2009/nov/03/youtube-funniest-comments.

Thibault, Paul J. 2000. "Multimodal Transcription of a Television Advertisement: Theory and Practice." In *Multimodality and Multimediality in the Distance Learning Age*, edited by Anthony P. Baldry, 311–85. Campobasso: Palladino.

Thibault, Paul. J. 2012. "Hypermedia Selves and Hypermodal Stories: Narrativity, Writing, and Normativity in Personal Blogs." In *Web Genres and Web Tools: Contributions from the Living Knowledge Project*, edited by Mariavita Cambria, Cristina Arizzi, and Francesca Coccetta, 7–50. Como and Pavia: Ibis.

Thurlow, Crispin, and Kristine Mroczec, eds. 2011. *Digital Discourse. Language in the New Media*. Oxford: Oxford University Press.

Tönnies, Ferdinand. 1957. *Community and Society*. New York: Harper Torchbooks.

Tottie, Gunnel. 1991. *Negation in English Speech and Writing. A Study in Variation*. San Diego, CA: Academic Press.

Tredinnick, Luke. 2006. *Digital Information Contexts: Theoretical Approaches to Understanding Digital Information*. Oxford: Chandos.

Tredinnick, Luke. 2008. *Digital Information Culture: The Individual and Society in the Digital Age*. Oxford: Chandos.

Tremayne, Mark. ed. 2006. *Blogging, Citizenship, and the Future of Media*. London and New York: Routledge.

Tribble, Chistopher. 1999. "Writing Difficult Texts." PhD dissertation. University of Lancaster.

van Lawick-Goodall, Hugo, and Jane van Lawick-Goodall. 1970. *Innocent Killers: A Fascinating Journey Through the Worlds of the Hyena, the Jackal, and the Wild Dog*. Boston: Houghton Mifflin.

van Leeuwen, Theo. 1999. *Speech, Music, Sound*. London: MacMillan.

van Leeuwen, Theo, and Carey Jewitt, eds. 2001. *The Handbook of Visual Analysis*. London: Sage.

Vansina, Jan. (1961) 2006. *Oral Tradition: A Study in Historical Methodology*. Translated by H. M. Wright. Harmondsworth: Penguin.

Ventola, Eija. 1979. "The Structure of Casual Conversation in English." *Journal of Pragmatics* 3: 267–98.

Vertegaal, Roel, and Yaping Ding. 2002. "Explaining Effects of Eye Gaze on Mediated Group Conversations: Amount or Synchronization?" In Proceedings of the 2002 ACM Conference on Computer Supported Cooperative Work, 41–8.

Vonderau, Patrick. 2009. "Writers Becoming Users: YouTube Hype and the Writer's Strike." In *The YouTube Reader*, edited by Pelle Snickars and Patrick Vonderau, 108–25. Stockholm: The National Library of Sweden.

Walker Rettberg, Jill. 2008. *Blogging. Digital Media and Society*. Cambridge and Malden: Polity Press.

Welch, Matt. 2003. "Blogworld and Its Gravity." *Columbia Journalism Review* 42(3), 21–7.

Wesch, Michael. 2009a. "From Knowledegeable to Knowledge-able: Learning in New Media Environments." *Academic Commons*. http://www.academiccommons.org/c

Wesch, Michael. 2009b. *The Machine is (Changing) Us: YouTube and the Politics of Authenticity*, Accessed November 26, 2010. http://www.youtube.com/watch?v=09gR6VPVrpw.

Winer, Dave. 2003. "What Makes a Weblog a Weblog?" Accessed March 20, 2012. http://web.archive.org/web/20040401132705/http://blogs.law.harvard.edu/whatMakesAWeblogAWeblog

Wray, Alison. 2002. *Formulaic Language and the Lexicon*. Cambridge: Cambridge University Press.

Wright, Laura, ed. 2006. *The Development of Standard English. 1300–1800. Theories, Descriptions, Conflicts*. Cambridge: Cambridge University Press.

Xiao, Richard. 2009. "Multidimensional Analysis and the Study of World Englishes." *World English* 28(4): 421–50.

Yakura, Elaine K. 2004. "'Informed Consent' and Other Ethical Conundrums in Videotaping Interactions." In *Discourse and Technology: Multimodal Discourse Analysis*, edited by Philip LeVine and Ronald Scollon, 146–50. Washington, DC: Georgetown University Press.

Yang, Ruigang, and Zhengyou Zhang. 2001. "Eye Gaze Correction with Stereovision for Video-Teleconferencing." Technical Report MSR-TR-2001–119. Accessed October 15, 2010. http://research.microsoft.com/en-us/um/people/zhang/Papers/TR01-119.pdf.

Yates, Simeon J. 1996. "Oral and Written Linguistic Aspects of Computer Conferencing. A Corpus-Based Study." In *Computer-Mediated Communication: Linguistic, Social and Cross-Cultural Perspectives*, edited by Susan C. Herring, 29–46. Amsterdam and Philadelphia: John Benjamins.

Yule, George. 2010. *The Study of Language*. 4th Edition. Cambridge: Cambridge University Press.

Zumthor, Paul. (1983) 1990. *Oral Poetry: An Introduction*. Translated by Kathy Murphy-Judy. Minneapolis: University of Minnesota Press.

Zumthor, Paul. 1984. "The Text and the Voice." *New Literary History* 16: 67–91.

Index

For Product Safety Concerns and Information please contact our EU
representative GPSR@taylorandfrancis.com
Taylor & Francis Verlag GmbH, Kaufingerstraße 24, 80331 München, Germany